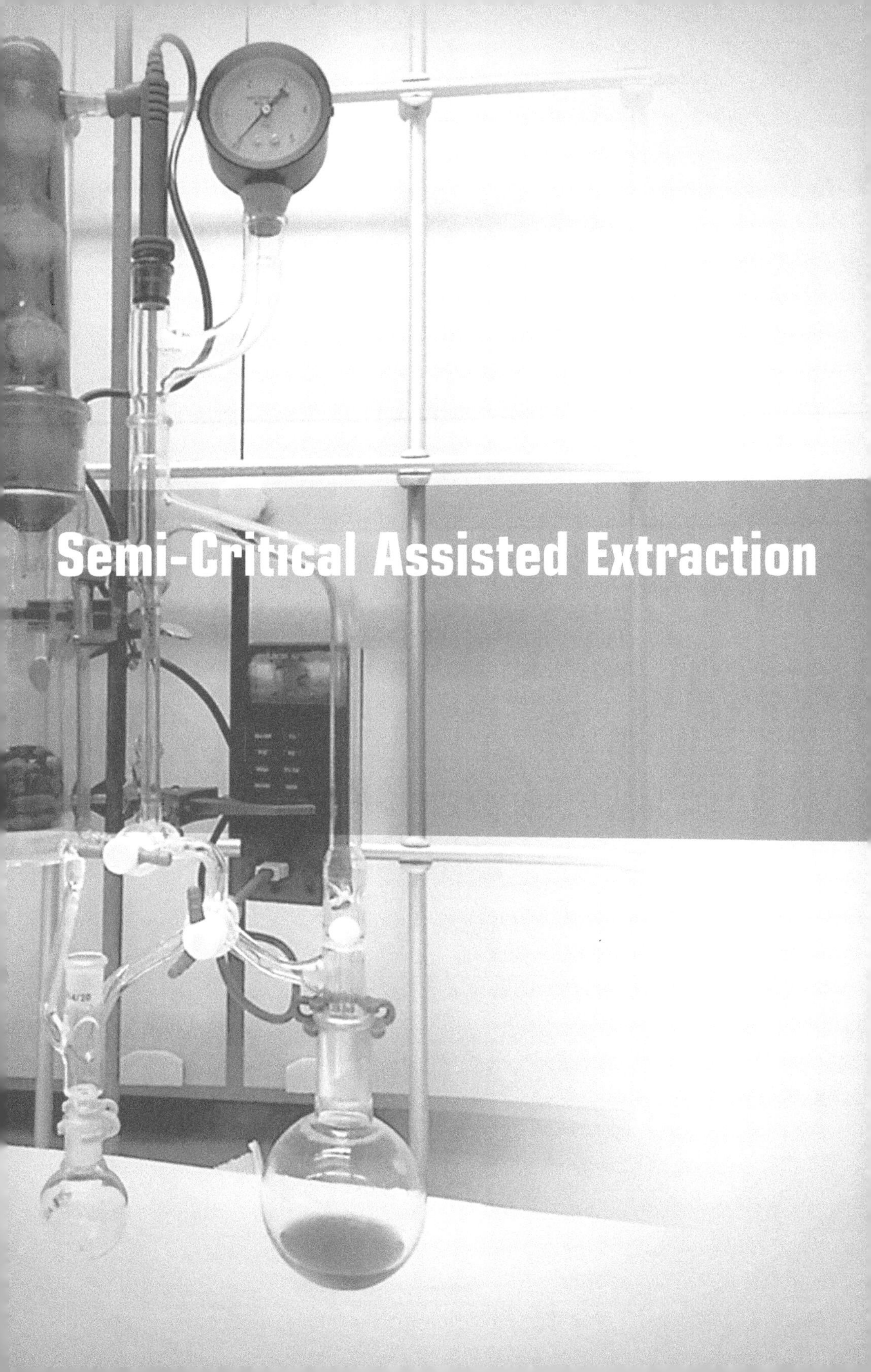

Semi-Critical Assisted Extraction

Semi-Critical Assisted Extraction

Applications and Commercialization in Biotechnology, Food, and Pharmacy

Tulio Chavez-Gil

Jenny Stanford Publishing

Published by

Jenny Stanford Publishing Pte. Ltd.
101 Thomson Road
#06-01, United Square
Singapore 307591

Email: editorial@jennystanford.com
Web: www.jennystanford.com

British Library Cataloguing-in-Publication Data
A catalogue record for this book is available from the British Library.

Semi-Critical Assisted Extraction: Applications and Commercialization in Biotechnology, Food, and Pharmacy

ISBN 978-981-4968-36-2 (Hardcover)
ISBN 978-1-003-29124-4 (eBook)

Contents

Preface

In the summer of 2011 and after a fruitful professional discussion on the advancements and limitations encountered for an engineer colleague on the commercial availability of continuous extraction equipment for algae oil extraction/conversion, I thought from that conversation that at that point, non-yet Soxhlet, Soxtec, Gregar, or current advanced glassware extraction technologies had been designed for in situ determination of key thermophysical variables such as pressure (P), temperature (T), and pH, occurring along a solvent extraction or during a chemical conversion process. Instead of that, temperature and pressure are set up as fixed parameters to perform extraction or separation when either one of the above-mentioned technologies is used to carry out an extraction experiment. Despite improvements in equipment designs used in either standard or advanced extraction technologies, an innovation that integrates these methods for in situ determination of thermophysical variables has been elusive in many ways to afford for the determination of these parameters. With the aim to overcome these limitations, a versatile compact extractor/separator apparatus was recently designed, constructed, tested, and USPTO patented. The innovation can be designed (changed) with special features allowing to perform conventional and semi-critical assisted extraction regarding thermodynamic adjustments to achieve the process needs. Phytochemistry, biochemistry, thermochemical, steady-state chemical extraction, molecular separation as well as conversion of systems required for food, pharmaceuticals, nutraceuticals, fragrances, and biotechnology are possible to be processed on the upgraded innovation.

Though it is a multi-functional design, the equipment has been tested under semi-critical assisted conditions as a technology that could provide the basis for a variety of extraction assessment to evaluate the current research extraction capability, analyze the progress of either chemical extraction or physical separation, compare its performance with more advanced extraction designs, and analyze challenges associated with current similar products for industrial scale-up with an insight into the expected forthcoming developments in advanced glassware equipment for extraction by applying biotechnology processes.

The goals of this book are to provide a realistic physical chemistry understanding of the principles related with thermochemical assisted extraction/conversion methods and how these principles can be applied in addressing socioeconomic and environmental impacts of the value added of chemical bioproducts derived from agriculture, food decontamination and processing, pharmaceutical bioproducts and analysis, as well as a valuable instrument for the assessment of established technologies in its possibilities and limitations for industrial scaling up.

The book also explores a new technology for the extraction/separation/conversion of raw materials along with the value added for new bioproducts, while discussing its development and practical improvements as an advanced technology with the potential to be integrated into biotechnology processes with application for assisted extraction or separation of dairy foods, egg food (nutritional analysis), edible oils, vitamins, proteins, and antioxidants by explaining its practical impact on yield, quality, and scale production.

A review of the current extraction technologies and their development and research findings is presented with an emphasis on the most recent breakthroughs in extraction, separation, and conversion technologies for the production and analysis of bioproducts.

The thematic I have attempted to present herein ensures a rigorous analysis and a readable comprehension, without the need for a structured and conceived knowledge from students or freshmen readers, whose primary interests are not necessarily chemistry, biotechnology, or engineering. The technical language I intend to use for describing critical concepts, mechanisms, procedures, data analysis, and special remarks is clear enough for those who are not chemistry or engineering majors, but have also contained in depth content required to include the book in an advanced graduate course.

Tulio Chavez-Gil

Ellicott City, MD
March 2023

Acknowledgments

Thanks, first to my wife, Falcohnery, and sons Tulio Enrique and Santiago for their understanding, support, and love through long hours over the many months of uncertainties related with the "COVID-19" pandemic threat.

Special thanks to my former students Mr. Alexis Pacheco-Laracuente and Mr. Sigfredo Villarin Ayala who collected and developed the first experiments with *Spirogyra* sp. macro algae; Mr. Pedro Rivera who worked on white egg-yolk at Inter American University of Puerto Rico, San German, PR; and Ms. Shante Lee at Coppin State University on testing the Ch-G extractor for lipids and protein extraction. I also appreciate the helpful discussions on Spent Coffee Ground and Micro Algae lipid extraction methods with Dr. Miriam Fontalvo (Dean of School of Pharmacy and Biochemistry, University of Atlántico, Colombia, SA) because after those discussions, the idea on the design and construction of a new extractor was conceived.

I greatly appreciate the GC-MS data collected by my colleague Dr. Hany F. Sobhi and the microscopy (micrographs on Lactose) -FESEM, EDS, TEM, and AFM recorded by Dr. Hyeonggon Kang at Coppin State University. Special thanks are offered to the University System of Maryland for the fantastic analytical instrumentation provided to our Department of Natural Sciences on that the data collected on proteins, amino acids, lipids, FAME, 2D, and 3D structure of proteins were recorded. Finally, I greatly appreciate the helpful suggestions of the reviewers at Jenny Stanford Publishing, who pointed out needed improvements in many ways on the manuscript body. Nonetheless, constraints on the scope and length of the book content mean that not all suggestions could be included, but I am truly thankful to all those that contribute to this first publication on the semi-critical extraction field.

Chapter 1

Fundamental Dimensions and Units for Extraction and Separation

1.1 Introduction

Independent of a science or engineering activity, a measurable interaction between energy and matter is always involved. Thus, it is hard to conceive an area that involves dynamic activity, which does not relate to thermodynamic analysis if one of these two physical principles (energy and matter) is present and interacts in some tangible measure. In that case, developing a good understanding of the physical, chemical, or thermodynamic principle that applies for a scientific method or industrialized process has long been an essential part in engineering, technology, as well as education. Thus, every physical quantity that implicitly measures a quantifiable variable or change during a process can be characterized by its intrinsic dimensions. Therefore, the arbitrary magnitudes assigned to the dimensions under analysis along any process are called their units.

Some basic dimensions such as length (L), mass (m), time (t), and temperature (T) are classified as primary or fundamental dimensions, while others such as velocity (V), energy (E), and volume (V) are expressed in terms of the primary dimensions and are called secondary dimensions, or "derived dimensions."

Semi-Critical Assisted Extraction: Applications and Commercialization in Biotechnology, Food, and Pharmacy
Tulio Chavez-Gil

ISBN 978-981-4968-36-2 (Hardcover), 978-1-003-29124-4 (eBook)
www.jennystanford.com

The systematic efforts to develop a universally acceptable system of units date back to 1790 when the French National Assembly charged the French Academy of Sciences to come up with such a unit system. An early version of the metric system was quickly developed in France at that time; however, it only found international acceptance after The Metric Convention Treaty of 1875 prepared a document signed by 17 countries, including the United States. In this first international treaty, meter and gram were established as the metric units for length and mass, respectively, and a General Conference of Weights and Measures (CGPM) was created and established to meet every 6 years. During the meeting of 1960, the CGPM produced the International System of Units (SI) document, which was based on six fundamental quantities and their units adopted in the meeting held in 1954 at the Tenth General Conference of Weights and Measures: (1) meter (m) for length, (2) kilogram (kg) for mass, (3) second (s) for time, (4) ampere (A) for electrical current, (5) degree Kelvin (K) for temperature, and (6) candela (cd) for luminous intensity (amount of light). In 1970, the CGPM added a seventh fundamental quantity and unit, (7) mole (mol), for the amount of matter. Those quantities, their units and symbols are shown in Table 1.1. In the notational scheme introduced in the CGPM 1967 meeting, the degree symbol was officially dropped from the absolute temperature unit, with the rest of unit names were to be written without capitalization even if they were derived from proper names. Nonetheless, the abbreviation of a unit was recommended to be capitalized if the unit was derived from a proper name. For example, the SI unit of force, which is named after Sir Isaac Newton (1647–1723), is newton (not Newton), and it is abbreviated as N.

Table 1.1 Seven fundamental dimensions and their SI base units

Quantity	SI Unit	Symbol
Length	meter[a]	m
Mass	kilogram	kg
Time	second	s
Temperature	kelvin	K
Amount of matter	mole	mol
Electric current	ampere	A
Amount of light	candela	cd

[a]or metre

Also, the full name of a unit may be pluralized, but its abbreviation cannot. For example, the length of an object can be 2.5 m or 2.5 meters, but not 2.5 ms or 2.5 meter. Finally, no period is to be written in unit abbreviations unless they appear at the end of a sentence; for example, the proper abbreviation of meter is m but not (m.).

1.2 Foundations of Quantities, Units, and Symbols

1.2.1 Physics and Chemistry Units

As stated earlier, there is an international agreement that the units used to express physical or chemical quantities in science, engineering, and technology may be those recommended by the International System of Units, or SI (standing for the French Système International d'Unités). A majority of physical chemistry quantities and units are denoted by using Arabic numbers and either a Greek or Latin symbol, respectively. With a few exceptions, this textbook will use symbols recommended for what is known as the IUPAC Green Book [1]. In this publication, the IUPAC manual states the use of symbols and scientific terminology based on the SI common units and it is produced by the International Union of Pure and Applied Chemistry (IUPAC) organization. The symbol for every physical or chemical quantity is given after a numerical quantity or within a parenthesis for convenience or clarity. Table 1.1 lists the seven SI base units, which will apply for all examples and problems in the book. These base units are for seven independent physical quantities that are sufficient to describe all other physical quantities.

1.2.2 The International System, SI, Revision of 2018

During the 26th General Conference on Weights and Measures (CGPM) that took place in Versailles, France, from 13 to 16 November 2018, a convocation of delegates representing more than 60 countries voted to implement the most significant change to the International System of Units (SI) in more than 130 years. In the meeting and for the first time, it was agreed that all measurement units will be defined by a natural phenomenon rather than by physical artifacts.

After decades of groundbreaking scientific work by metrologists at the national measurement institutes (NMIs) from around the world, the delegates voted to redefine the kilogram and other three additional basic SI units: electric current (ampere), temperature (kelvin), and amount of substance (mole). Those new definitions became effective on May 20, 2019, declared as the "World Metrology Day," which celebrates the establishment of the SI, or metric system, in 1875. In this meeting, the base units for mass, thermodynamic temperature, amount of substance, and electric current were redefined. The SI revision bases the definitions of the base units (Table 1.2) on a set of six defining constants with values listed in Table 1.3 treated as exact, with no uncertainty. Thus, the kilogram had been previously defined as the mass of a physical artifact, an accepted international prototype of the kilogram. This international prototype is a platinum–iridium cylinder manufactured in 1879 in England and stored since 1889 in a vault of the International Bureau of Weights and Measures in Sèvres, near Paris, France. As it is a material subject to surface contamination and other slow changes of mass, it is not entirely suitable at this time to be accepted as a standard. To ensure some solutions to these discrepancies, in the 2019 SI revision instead, the commission redefined the kilogram in terms of the Planck constant $\hbar$ [2]. As a defining constant, the value of $\hbar$ was chosen to agree with the mass of the international prototype with an uncertainty of only several parts in 108. Thus, as a practical matter, the SI revision has a negligible effect on the value of a mass. The SI revision defines the kelvin in terms of the Boltzmann constant k, the mole in terms of the Avogadro constant N_A, and the ampere in terms of the elementary charge *e*. The values of these defining constants were chosen to closely agree with the previous base unit definitions. Therefore, the 2018 SI revision has a negligible effect on the values of thermodynamic temperature, amount of substance, and electric current physical properties.

1.2.3 Standard Prefixes in SI

As pointed out earlier, the SI is based on a decimal relationship between units and to express the multiples of the various units, the SI recommends the usage of prefixes, as seen in Table 1.4. They are standard to all units, and the reader is encouraged to take them in mind (memorize) because of their widespread use.

Table 1.2 SI-derived units

Physical Quantity	Unit	Symbol	Definition of Unit
Force	newton	N	$1\ N = 1\ m\ kg\ s^{-2}$
Pressure	pascal	Pa	$1\ Pa = 1\ N\ m^{-2} = kg\ m^{-1}s^{-2}$
Celsius temperature	Celsius	°C	$t/°C = T/K - 273.15$
Energy	joule	J	$1\ J = 1\ N\ m = 1\ m^2\ kg\ s^{-2}$
Power	watt	W	$1\ W = 1\ J\ s^{-1} = 1\ m^2\ kg\ s^{-1}$
Frequency	hertz	Hz	$1\ Hz = 1\ s^{-1}$
Electrical charge	coulomb	C	$1\ C = 1\ A\ s$
Electric potential	volt	V	$1\ V = 1\ J\ C^{-1} = 1\ m^2\ kg\ s^{-3}\ A^{-1}$
Electric resistance	ohm	W	$1\ W = 1\ V\ A^{-1} = m^2\ kg\ s^{-3}\ A^{-2}$

Table 1.3 Non-SI-derived units

Physical Quantity	Unit	Symbol	Definition of Unit
Volume	liter[a]	L[b]	$1\ L = 1\ dm^3 = 10^{-3}\ m^3$
Pressure	bar	bar	$1\ bar = 10^5\ Pa$
Pressure	atmosphere	atm	1 atm = 101,325 Pa = 1.01325 bar
Pressure	torr	Torr	1 Torr = (1/760) atm
Pressure	torr	Torr	1 Torr = (101,325/760) Pa
Energy	calorie[c]	Cal[d]	1 Cal = 4.184 J

or [a]litre, or [b]l, or [c]thermochemical calorie, or [d]cal^{th}

1.2.4 Scientific Notation

Scientists often work with numbers that are extremely large or extremely small. For example, there are 10,300,000,000,000,000,000,000 carbon atoms in a 1-carat diamond gemstone each of which has a mass of 0.000,000,000,000,000,000,0 00,020 grams. Therefore, it is impossible in current days to multiply these numbers with either a scientific or graphic calculator because they cannot accept either number as it is written here.

Table. 1.4 Standard prefixes in SI units

Multiple	Prefix	Symbol	
10^{12}	tera	T	
10^{9}	giga	G	Prefixes that increase size
10^{6}	mega	M	
10^{3}	kilo	k	
10^{-1}	deci	*d*	
10^{-2}	centi	*c*	
10^{-3}	milli	*m*	
10^{-6}	micro	μ	Prefixes that decrease size
10^{-9}	nano	*n*	
10^{-10}	angstrom	Å*	
10^{-12}	pico	*p*	
10^{-15}	femto	*fs*	

*Non-SI unit but accepted worldwide in sciences and engineering.

To successfully get a result by doing a calculation like this, it is necessary to express these numbers simply in **scientific notation**, as a number between 1 and 10 multiplied by 10 raised to some exponent, like 10^{n}.

Problem 1.1

Convert the following standard numbers into scientific notation.

(a) 0.005793 *Answer*: 5.793×10^{-3}
(b) 35.2 *Answer*: 3.52×10^{1}
(c) 2,486,000 *Answer*: 2.486×10^{6}

1.2.5 Rounding Off and Significant Figures

It is the packager's responsibility to round converted values appropriately and select the appropriate number of significant digits to use in a quantity declaration. These rounding rules are for converting quantity determinations on packages and do not apply to digital scales that automatically round indications to the nearest indicated value.

Table 1.5 Scientific notation numbers

Number	Multiples of 10	Scientific Notation	
10 0000	$10 \times 10 \times 10 \times 10 \times 10$	1×10^5	
10 000	$10 \times 10 \times 10 \times 10$	1×10^4	
1000	$10 \times 10 \times 10$	1×10^3	Positive powers of 10
100	10×10	1×10^2	
10	1×10	1×10^1	
1	0	1×10^0	
0.1	$\frac{1}{10}$	1×10^{-1}	
0.01	$\frac{1}{10} \times \frac{1}{10} = \frac{1}{100}$	1×10^{-2}	
0.001	$\frac{1}{10} \times \frac{1}{10} \times \frac{1}{10} = \frac{1}{1000}$	1×10^{-3}	Negative powers of 10
0.0001	$\frac{1}{10} \times \frac{1}{10} \times \frac{1}{10} \times \frac{1}{10} = \frac{1}{10000}$	1×10^{-4}	
0.00001	$\frac{1}{10} \times \frac{1}{10} \times \frac{1}{10} \times \frac{1}{10} \times \frac{1}{10} = \frac{1}{100000}$	1×10^{-5}	

1.2.6 Conversions

The proper use of significant digits and rounding off must be based on the packer's knowledge of the accuracy of the original measurement that is being converted. For example, if a package is labeled 453.59 g (1 lb), the packer is implying that the package declaration is accurate within ±0.005 g (or ±5 mg). Similarly, for liquid volume expressions, a label declaration of 473 mL (16 fl oz) implies that the package declaration is accurate to within ±0.5 mL (0.01 fl oz).

1.2.7 Rounding

In all conversions for the purpose of showing an equivalent SI or US customary quantity to a rounded US customary or SI quantity, or in calculated values to be declared in the net quantity statement, the number of significant digits retained must be such that accuracy is neither sacrificed nor exaggerated. Conversions, the proper use of significant digits, and rounding must be based on the packer's knowledge of the accuracy of the original measurement that is being converted. In no case shall rounded net contents declarations overstate a quantity; the packer may round the converted values down to avoid overstating the net contents.

NOTE: *When, as a result of rounding SI or US customary declarations calculated based on the conversion factors in Appendix A, the resulting declarations are not exact; the largest declaration, whether metric or US customary, will be used for enforcement purposes to determine whether a package contains at least the declared amount of the product.*

Do not round conversion factors or any other quantity used or determined in the calculation; one only rounds the final quantity to the number of significant digits needed to maintain the accuracy of the original quantity. Use the rounding rules presented in Table 1.6 as guidance to round the final result. In general, quantity declarations on consumer commodities should only be shown to two or three significant digits (for example, 453 g or 85 g). Any final zeros to the right on the decimal point need not be expressed. The US customary and SI declarations of quantity must be accurate and equivalent to each other. For example, a package bearing a net weight declaration of 2 lb (32 oz) must also include an SI declaration of 907 g.

In any calculation, the answer must be expressed by possessing the same number of significant figures as the measured numbers. Thus, calculator answers must often be **rounded off** by following simple but practical rules that make calculations more accurate or precise.

1.2.8 Rounding Rules

These rules are used to obtain the correct number of significant figures, depending on the mathematical operation that applies for the analysis carried out. For example, when the first digit dropped is **4 or less than 5**, the retained numbers remain the same. Thus, to round 45.832 to three significant figures, it is necessary to drop the last two digits 3 and 2 to obtain the answer as 45.8. Otherwise, when the first digit dropped is **5 or greater**, the last retained digit is increased by 1. Hence, to round 2.4884 to two significant figures, it is necessary to drop the digits 8, 8, and 4 to get the answer as 2.5 (herein, 0.4 was increased by 0.1).

Table 1.6 Significant figure rules for measured and calculation numbers

		Measured	Number of Significant
Rule	**Definition**	**Number**	**Figures**
1.	***A number is significant if it is:***		
	a. Not a zero	3.5 s	2
		123.34 m	5
	b. A zero between digits	2405 cm	4
		102.023 km	6
		1203.2004 mm	8
	c. A zero at the end of a decimal number		
		300. g	3
		43.0 s	3
		32.000 lb	5
		98.700 °C	5
	d. Any digit in the coefficient in a number written as scientific notation		
		2.0×10^4 m	2
		6.023×10^{23} atoms	4
2.	***A zero is not significant if it is:***		
	a. At the beginning of decimal number		
		0.000032 g	2
		0.00753 nm	3
	b. Used as a placeholder in a large number without a decimal point		
		234560000 m	5
		2 345 000 000 kg	4

1.2.8.1 Multiplication and division in rounding off

When **multiplying** or **dividing**, use the same number of significant figures (SF) as the measurement with the fewest significant figures.

Example 1.1: Round the follow ***multiplication***.

$$\underset{\text{4SFs}}{110.5} \times \underset{\text{2SFs}}{0.048} = \underset{\textit{calculator}}{5.304} = \underset{\text{answer 2SFs}}{5.3 \text{ (rounded)}}$$

Example 1.2: Round the follow ***division***.

$$\underset{\text{4SFs}}{5.311} \div \underset{\text{2SFs}}{0.15} = \underset{\textit{calculator}}{35.40666667} = \underset{\text{answer 2SFs}}{35 \text{ (rounded)}}$$

1.2.8.2 Addition and subtraction in rounding off

When adding or subtracting, use the same number of decimal places that are present in the numeral with the fewest decimal places by following the rounding rules to adjust the number of digits in the answer.

Example 1.3: Round the follow ***addition***.

35.**2**	*one decimal place*
+ 3.**24**	*two decimal places*
38.**44**	*calculated answer*
38.4	*final answer (adjusted to one decimal place)*

Example 1.4: Round the follow ***subtraction***.

$$68.72539 - 33.08 = \underbrace{35.64539}_{\substack{\textit{Five}\\ \textit{decimal}\\ \textit{places}}} = \underbrace{35.65}_{\substack{\textit{Two}\\ \textit{decimal}\\ \textit{places}}} \text{ final answer (adjusted to two decimal places)}$$

Calculated answer

1.2.9 Significant Figures

In every research or industrial process, either a quantitative or a qualitative analysis is reported as part of a physical, chemical, biochemical, or biotechnological activity that involves derived properties of the obtained substances or as key parameters that render significant information from all measurements performed. The measured results must include all the known digits plus the estimated digit(s). In addition, the number of significant figures reported in a measurement will depend on the accuracy or calibration of the measuring tool used in that measurement. Nevertheless, some rules are necessary to take in mind when significant figures are used to report a measurement.

1.3 Amount of a Substance: The SI 2018 New Definition

In science or engineering fields, "*quantity*" is formally understood as the amount of substance, which is the counting of a quantity for a specified elementary entity. An elementary entity can also be described as an atom, a molecule, an ion, an electron, a proton, a nuclide, or any other particle or specified group of particles. The SI base unit for the amount of substance is defined as "the mole" [3]. Before 2018, the mole was defined as the amount of substance containing as many elementary entities as the number of atoms in exactly 12 g of pure carbon-12 nuclide, ^{12}C. This definition was such that 1 mol of CO_2 molecules, for example, has a mass of 14.03 g, where 14.03 is the relative molecular mass of 1 mol of CO_2, which contains 6.022×10^{23} molecules of CO_2, where 6.022×10^{23} mol^{-1} is N_A, the Avogadro's constant (values given to four significant figures). The same statement can be made for any other substance if 14.03 is replaced by the appropriate relative atomic mass or molecular mass value. In the SI revision of 2018 (section 1.2.2), it **redefines** the mole as exactly $6.022{,}140{,}76 \times 10^{23}$ elementary entities. Thus, the mass of this number of carbon-12 atom is 12 g to within 5.0×10^{-9} g [4], and so the revision makes a negligible change in calculations involving the mole unit. The symbol for an amount of substance is expressed as a coefficient n (the mole symbol), and it is written before the chemical formula. Therefore, it is admittedly awkward to refer to n (CO_2) instead of $(CO_2)_n$ as "the amount of substance of carbon dioxide" in terms of moles. In this book, we simply shorten "amount of substance" to "amount of," a usage condoned by the IUPAC to write any formulae name [4]. Thus, "the amount of carbon dioxide in the system" refers not to the mass or volume of CO_2, but specifically to the number of CO_2 molecules expressed in a counting unit such as the number of *moles* of CO_2 in either chemical formula or reaction.

1.4 Quantity Calculations in Science and Technology

Following SI changes and new definitions, this section examines how one may manipulate either a physical or a chemical quantity under the rules of algebra [5]. The general mathematical method for the purpose of this book is the so-called *quantity calculus*, although the term that better defines a physical or chemical calculation herein will be defined as *algebraic operations*, which operates on real numbers and real scalar variables. The quantity calculus method is based on the concept that a physical quantity, unless it is dimensionless, has a value equal to the product of a numerical value (a real number) and one or more units associated with the variable under analysis:

1. Physics

$$\text{Physical quantity} = \text{Real number} \times \text{Units} \tag{1.1}$$

2. Chemistry

$$\text{Chemical quantity} = \text{A coefficient (real number)} \times \text{Chemical formulae (atom(s), molecule(s))} \tag{1.2}$$

If the quantity is dimensionless, it is equal to a pure number without units, as for example in calculating the specific gravity, SG of a substance X, where the calculation is expressed as:

$$\text{SG} = \frac{\text{Sample's density}}{\text{Water density}} = \frac{X\,\text{g/cm}^3}{1\,\text{g/cm}^3} = X, \quad \text{a real number without units} \tag{1.3}$$

On the other hand, when a physical state is the subject of analysis, as happens during a chemical reaction, the states of the substances in the reaction are denoted by a subscript symbol, such as *s* (for solid), *l* (for liquid), *g* (for gas), *aq* (for aqueous), or *sol* (for solution), but the symbol does not imply a particular choice of units because it is related specifically with the state of matter only.

To analyze the mathematical relationships between physical dimensions and properties expressed for any kind of matter, this book uses the thermodynamics symbol H, for its specific heat.

Nonetheless, H can be expressed in different units having the dimensions of 4.186 J divided by grams × °C (J/g·°C), or calories divided by grams × °C (cal/g·°C); according to the area of science, engineering, or technology, the property is under analysis. The following example illustrates the use of quantity calculus expressed in terms of joules and/or calorie units. Herein, we can express the specific heat of water, H_w, in joules (J) or calories (Cal) at 25°C to four significant figures in SI base units by the following equations:

$$H_w = \frac{4.184 \times 10^3 \text{J}}{\text{kg°C}} = 4184 \text{ kJ} \tag{1.4}$$

or

$$H_w = \frac{4.184\text{J}}{\text{g°C}} = 4.184 \text{ J} \tag{1.5}$$

also,

$$H_w = \frac{1.000 \times 10^3 \text{Cal}}{\text{kg°C}} = 1 \text{ kCal} \tag{1.6}$$

or

$$H_w = \frac{1.000 \text{ Cal}}{\text{g°C}} = 1 \text{ Cal} \tag{1.7}$$

and

$$H_w = \frac{1 \text{ Cal}}{\text{g°C}} = 4.184 \frac{\text{J}}{\text{g°C}} \tag{1.8}$$

Similarly, it can be expressed as an exact number but in different specific heat units by the following equation:

$$C_\text{p} = \frac{\text{BTU}}{\text{lb m } o_\text{F}} \tag{1.8a}$$

With C_p, specific heat of water, BTU are British thermal units, which is defined as the amount of heat required to raise the temperature of one pound (m, mass) of water by one-degree Fahrenheit, at the temperature water has its greatest density (~39 degrees Fahrenheit = 3.88 °C), very different to the SI definition that uses one gram of water (not one pound) and $\overline{1}$.00 °C to determine the same physical property. Similarly, 1.00 BTU is about 1055 J or 252.05597 calories. These quantities can be used to substitute the SI common values on the above equations on our analysis for the mathematical relationships between water physical dimensions and properties when expressed for any other kind of matter analysis.

Herein, if we divide both sides of Eqs. 1.5 and 1.7 by J g^{-1}°C^{-1} and Cal g^{-1} °C^{-1}, respectively, one obtains the following:

$$\frac{H_w}{\text{J/g°C}} = \frac{4.184 \text{ J/g°C}}{\text{J/g°C}} = 4.184 \tag{1.9}$$

Now the pure number 4.184 appearing in Eq. 1.9 is the number of joules content in 1 g of water at 25°C, and so we may call the ratio joules/g°C "the number of joules per gram per Celsius degrees" in terms of the energy content in a specific mass of matter and at that temperature. By the same reasoning,

$$\frac{\frac{H_g}{\text{Cal}}}{\text{g°C}} = \frac{\frac{1\text{ Cal}}{\text{g°C}}}{\frac{1\text{ Cal}}{\text{g°C}}} = 1.000 \tag{1.10}$$

where 1.000 is the pure number of calories per gram of water in the same number of Celsius degrees measured in Eq. 1.9. Therefore, a physical quantity divided by same particular units for the physical quantity under analysis gives a pure number that represents the same number of those units as defined for the SI for the physical dimension used in that calculation. Performing algebraic calculations by using the SI base units together with its derived units results in real numbers, which give coherence for any physical or chemical property evaluation. Thus, the values of a physical or chemical quantity expressed in different mathematical combinations with its respective units have the same numerical value when algebraic calculations are properly equalized as proposed in the following example:

Example 1.5:
Pressure, P, is a physical and chemical property exerted by liquids and gases; therefore, both fields use the ideal gas equation to evaluate the pressure of a gas using the same units:

$$P = \frac{nRT}{V} \tag{1.11}$$

In this equation, p, n, T, and V are the symbols for the physical quantities pressure, amount of substance, thermodynamic temperature, and volume, and R is the gas constant (= 8.3145 $\text{J·K}^{-1}\text{·mol}^{-1}$). However, R = 0.0821 62.4000 $\text{mmHg} \cdot \text{L} \cdot \text{mol}^{-1} \cdot \text{K}^{-1}$ if the pressure units are mmHg instead of atmospheres. Therefore, the thermodynamic calculation of pressure p for 0.020000 moles of an ideal gas at room temperature (= 298.15 kelvins), in a volume of 15.000 cubic meters, is

$$P = \frac{(0.02000\text{ mol}) \times (8.3145\text{ J/K}^{-1} \cdot \text{mol}^{-1}) \times (298.15\text{ K})}{15.000\text{ m}^3} = \frac{3.305\text{ J}}{\text{m}^3} \tag{1.12}$$

In this equation, moles and kelvin units cancel, and we are left with J and m^{-3} units, which in a combination of SI-derived unit (the joule) and SI base unit (the meter) give the pascal units for pressure. Thus, the units J and m^{-3} must have dimensions of pressure, but they are not commonly used to express pressure in simple lab operations. Therefore, to convert joules and volume, J m^{-3} to the SI derived unit of pressure, the pascal (Pa), we can use the following mathematical relation of units as presented in Table 1.7.

$$1\ \text{J} = 1\ \text{N} \cdot \text{m} \quad \text{and} \quad 1\ \text{Pa} = 1\ \text{N} \cdot \text{m}^{-2} \tag{1.13}$$

Table 1.7 Values of the universal gas constant, *R*

$R = 0.0820578\ (\text{atm} \cdot \text{L})/(\text{mole} \cdot \text{K})$
$= 62.358\ (\text{torr} \cdot \text{L})/(\text{mole} \cdot \text{K})$
$= 62{,}358\ (\text{torr} \cdot \text{cm}^3)/(\text{mole} \cdot \text{K})$
$= 62{,}358\ (\text{mmHg} \cdot \text{cm}^3)/(\text{mole} \cdot \text{K})$
$= 0.083145\ (\text{bar} \cdot \text{L})/(\text{mole} \cdot \text{K})$
$= 8.\ 31451\ (\text{Pa} \cdot \text{m}^3)/(\text{mole} \cdot \text{K})$
$= 8.3145\ (\text{J})/(\text{mole} \cdot \text{K})$
$= 82.05\ (\text{atm} \cdot \text{cm}^3)/(\text{mole} \cdot \text{K})$
$= 8.205 \times 10^{-5}\ (\text{atm} \cdot \text{m}^3)/(\text{mole} \cdot \text{K})$
$= 0.00199\ \text{kcal}/(\text{mole} \cdot \text{K})$
$= 1.987\ \text{cal}/(\text{mole} \cdot \text{K})$

Adapted from Ref. [6].

Thus, if we divide both sides of the left equality by 1 J, and both sides of the right equality are divided by 1 N · m^2, there are obtained two new relations

$$1 = \left(\frac{1\ \text{N} \cdot \text{m}}{\text{J}}\right) \quad \text{and} \quad \left(\frac{1\ \text{Pa}}{1\ \text{N} \cdot \text{m}^{-2}}\right) = 1 \tag{1.14}$$

In the new equalities, the ratios in parentheses are thermodynamic conversion factors. Hence, when a physical quantity is multiplied by a conversion factor that, like these, the physical quantity changes its thermodynamic units but not its value. Therefore, by multiplying Eq. 1.14 by both these conversion factors, all units except Pa cancel:

$$P = (3.305 \times 10^3\ \text{J} \cdot \text{m}^{-3}) \times (1\ \text{N} \cdot \text{m}/\text{J}) \times (1\ \text{Pa}/\text{N} \cdot \text{m}^{-2}) = 3.305 \times 10^3\ \text{Pa} \tag{1.15}$$

With this example, we illustrate the fact that to calculate a physical chemistry quantity, we can simply enter in a calculator numerical values expressed in that system units. In other words, as long as we use only SI base units and SI-derived units (without prefixes), all conversion factors must conduce to unity. However, our calculations not necessarily need to be limited to SI units. For example, we could express the calculated pressure in ***torr*** units, a non-SI unit. In this case, using a conversion factor obtained from the definition of Torr in Table 1.7, the calculation becomes

$$P = (3.305 \times 10^3 \text{ Pa}) \times \left(\frac{760 \text{ torr}}{101{,}325 \text{ Pa}} \right) = 24.80 \text{ torr} \qquad (1.16)$$

1.5 Analysis of Dimensions

When performing an analysis of any kind of dimension, it implies that if a significant number of variables are acting on a given process, they are best analyzed if grouped into dimensionless parameters, which are less numerous than that set up or generated during the experimental process. Such a procedure is very helpful in experimental work as the number of significant variables can present an important task of correlation of units. Therefore, by doing a combination of variables with the purpose of cancel those units not desired in the results, we obtain a small number of dimensionless parameters which turns sometimes complex calculations into a reduced experimental data analysis.

1.5.1 Homogeneity in Dimensional Analysis

We all know from grade school that oranges and grapes do not add. But we somehow manage to do it (by mistake, of course). However, in science, engineering, and technology analysis, all equations must be *dimensionally homogeneous*. That is, each term in a physics or mathematical equation must have the same units for a manual or computed analysis. If at some stage of an equation or expression analysis, we find ourselves in a position to add two quantities that have different units, it is a clear indication that we had made an error at an earlier stage that need our attention by checking for accurate dimensional consistency. Therefore, checking data units can serve as a valuable tool to spot errors. A short but useful set of hints have been suggested as simple rules that must satisfy units correction in a physical equation or mathematical expression:

a. Both sides of one equation have the same dimensions.
b. All terms in an addition or subtraction operation have the same dimensions.
c. All logarithm or exponential and arguments of logarithms and exponentials are dimensionless.
d. Any power of 10 quantity used is dimensionless.
e. Assurance that significant figures and rounding off rules meet the final results.

Thus, we know from experience that units can give terrible headaches if they are not used carefully in solving a problem. But paying a little attention and using basic math skills, units can be used to benefit our needs. Units can be used to check the accuracy of physical and chemical formulae as well as to derive formulae, as explained in the following example:

Example 1.6:
An extractor cylinder is filled with a contaminated fuel oil whose density is $\rho = 750.0\ \text{kg/m}^3$ at 140°C. If the volume of extractor is $V = 1.2\ \text{m}^3$, determine the amount of fuel oil (m) in the extractor cylinder.

Solution: If we forgot the formula that relates mass to density and volume, but we know that mass has the units of kilogram (kg), and density has units of mass in function of volume, it is a good start. Thus, whatever calculation we performed with the data, we should end up with the units of kilograms. Therefore, putting the given information into mathematical perspective, we got

$$\rho = 750.0\ \text{kg/m}^3 \quad \text{and} \quad V = 1.2\ \text{m}^3$$

Fuel oil data for Example 1.6

$$\rho = 750\ \text{kg/m}^3$$

$$V = 10\ \text{m}^3$$

$$m = ?$$

It is obvious that from the units of volume (m^3), we can eliminate the m^3 units and end up with kg by multiplying these two quantities in a conversion factor math operation. Therefore, the formula we are looking for is

$$m = \rho \cdot V \tag{1.17}$$

Thus, $m = (750.0 \text{ kg/m}^3) \times (1.2 \text{ m}^3) = 900.0 \text{ kg}$

Therefore, we should keep in mind that a formula that is not dimensionally homogenous is definitely wrong; however, a dimensionally homogenous formula is not necessarily correct.

Problem 1.2

In a cold Alaska winter, a 10 kg plastic tank with a volume capacity of 0.5 m^3 was filled with ice. After ice melted, water was obtained. Assuming that water density is 1000.00 kg/m^3, determine the total weight for the last combined system.

Answer: 510 kg

1.6 Correlation in Experimental Dimensions

The practical extraction or separation of compounds by employing standard and/or advanced extraction technologies always involves the mass-transfer phenomenon mostly occurring in terms of turbulent flow. It is difficult to compute with some extent of accuracy the mass-transfer coefficients from theoretical considerations. To overcome these challenges, it is necessary to rely on collected enough experimental data. Hence, depending on the used mathematical system, the data could be limited in scope, but in certain circumstances, it can be obtained from the known range of the fluid properties. Thus, it is important to be able to extend the knowledge of applicability of coefficients to conditions not covered by the experiment parameters but that allow to draw upon our knowledge of other transport processes (heat, particularly), which can help in the required analysis. In this regard, a useful procedure is the so-called *dimensional analysis*.

Through this procedure, the significant variables operating on a given extraction process are grouped into dimensionless parameters, which are less numerous than those of the original variables. The procedure has proved to be very helpful in experimental work if a large number of variables represent a challenge for their correlation. In this analysis, a combination of variables is used to decrease from a larger to a smaller number of dimensionless parameters by making the experimental work data more accessible for calculations.

In general, dimensional analysis can be used to predict the various dimensionless parameters that are helpful in correlating experimental data to determine, for example, the mass-transfer coefficients in an extraction process. Nonetheless, certain dimensions may be established as fundamentals, while the others are expressed in terms of the reduced ones. Within those expressed as fundamental dimensions, we can mention *length*, symbolized as L. Thus, area A and volume V may dimensionally be expressed as L^2 and L^3, respectively. An implicit fundamental dimension that plays a key role in every scientific or technological process is **time**, symbolized t. For example, velocity and acceleration may be expressed as L/t and L/t^2 in a physics analysis, respectively. Another dimension to consider as fundamental in every research field is *mass*, symbolized as m. Also, the mol concept is included as physical chemistry definition of atomic and molecular quantities related with m. For example, a physical property whose dimensional expression relays with mass is density (mass or molar), which can be expressed as m/L^3. Similarly, chemical reactions require varying length of time for completion, which depends on the reactants and product characteristics among the conditions under which the reaction is carried out.

1.7 Rate of Reaction: Dimension's Measurement

The rate of a chemical reaction is the *chemical kinetics* of the reaction, which is referred to as the change in the concentration Δ[M] of a reactant or product(s) with change in time Δ[s]. The rate of a reaction (Fig. 1.1) is expressed in terms of two fundamental dimensions, M (moles of reactants or product) and t with the reaction occurring under constant physical conditions, k.

Hence,

$$\textit{rate} \text{ of formation/decomposition of compound A} = \pm \Delta[A]_t/\Delta t \quad (1.18)$$

This equation gives the *average* rate over the time interval Δt for the formation or decomposition of compound A. The ± sign relays with physical formation (+) or decomposition (–) interpretation. Thus, the chemical reaction rate expresses a change in concentration for a reactant or a product during a time interval ($\Delta M/\Delta s$) before it reaches the equilibrium; therefore,

$$\text{Reactant(s)} \rightarrow \text{Product(s)}$$

$$\text{nA} \rightarrow \text{mB}$$

$$\text{Product average rate} = \frac{\text{Change \# moles of B}}{\text{Change in time}} = \frac{\Delta\,(\text{moles B})}{\Delta t} = +\frac{\Delta\,[\text{B}]}{\Delta t} \tag{1.19}$$

while reactants go away, reactant's rate average is

$$\text{Rate} = -\frac{\Delta\,[\text{A}]}{\Delta t} \tag{1.20}$$

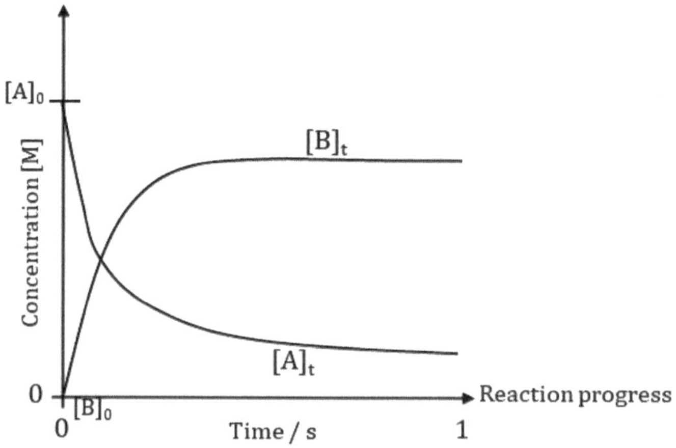

Figure 1.1 Rate of a chemical reaction.

The decomposition of dinitrogen pentoxide (N_2O_5) occurs at medium temperatures to give NO_2 plus O_2.

Calculate the average rate for the decomposition of N_2O_5, $-\Delta[N_2O_5]$; and the average rate of formation for NO_2, $\Delta[NO_2]$, and O_2 $\Delta[O_2]$, respectively; if the decomposition reaction is

$$2N_2O_5 \rightarrow 4NO_2 + O_2 \tag{1.21}$$

during the time interval from $t = 500$ s to 600 s with the experimental data shown in Table 1.8.

Table 1.8 Experimental data in the decomposition of N_2O_5 at medium temperature (20–25°C) and 1 atm

Time (s)	Concentration (M)		
	N_2O_5	N_2O	O_2
0	0.0200	0	0
100	0.0169	0.0063	0.0016
200	0.0142	0.0115	0.0029
300	0.0120	0.0160	0.0040
400	0.0101	0.0197	0.0049
500	0.0086	0.0229	0.0057
600	0.0072	0.0256	0.0064
700	0.0061	0.0278	0.0070

Data obtained from Ref. [7].

Problem setup: An average reaction rate is a change in the concentration of a reactant or product over the time interval. In this case, it is the rate of decomposition of the reactant N_2O_5 ($-\Delta[N_2O_5]/\Delta t$); and the product rate average for NO_2 ($\Delta[NO_2]/\Delta t$) and for O_2 ($\Delta[O_2]/\Delta t$) formation, respectively. Here, the reactant decomposition resolves as $-\Delta[A]/\Delta t$ in Eq. 1.20, which is the change in the concentration observed on the reactant (final value minus initial value). On the other hand, Δt corresponds to the time interval (final minus initial) over which the change in concentration occurred (herein, 600s – 500s). Thus, in a steady-state chemical reaction, the reactant's concentration always decreases during any time interval, but it is necessary to take into consideration if all the physical variables that apply for the reaction remain constant or the reaction parameters had to be setup to achieve these goals.

Average rate of decomposition of N_2O_5 = $-\Delta[N_2O_5]/\Delta t$

$$\text{Rate of } N_2O_5 = -\frac{(0.0072-0.086)\,\text{M}}{(600-500)\,\text{s}} = \frac{(-0.0014)\,\text{M}}{100\text{s}} = 1.4\times 10^{-5}\,\text{M/s}$$

Average rate of formation of NO_2 = $\Delta[NO_2]/\Delta t$

$$\text{Rate of } NO_2 = -\frac{(0.0072-0.086)\text{ M}}{(600-500)\text{ s}}$$

$$= \frac{(-0.0014)\text{ M}}{100\text{s}} = 1.4\times10^{-5}\text{ M/s}$$

Why do they differ?

Average rate of formation of O_2 = $\Delta[O_2]/\Delta t$

Recall the reaction is $2N_2O_5 \rightarrow 4NO_2 + O_2$

$$\text{Rate of } O_2 = \frac{(0.0064-0.0057)\text{ M}}{(600-500)\text{ s}}$$

$$= \frac{(0.0007)\text{ M}}{100\text{s}} = 7.0\times10^{-6}\text{ M/s}$$

From the above results, the rate average for N_2O_5 decomposition is twice of that calculated for O_2 formation in the same time interval (within the experimental error of the values given in the preceding discussion).

Thus, be sure the reaction rates are expressed as a positive number; otherwise, if the rate value is a negative quantity, it is likely because you did not treat D quantities as a final value minus an initial value. Therefore, to compare all reaction rates and make them dimensionally equivalents, it is necessary to account for the equation's ***stoichiometry***.

Thus,

N_2O_5 = **1/2** $\times 1.4\times10^{-5}$ M/s = 7.0×10^{-6} M/s
NO_5 = **1/4** $\times 2.7\times10^{-5}$ M/s = 6.75×10^{-6} M/s
O_2 = **1/1** $\times 7.0\times10^{-6}$ M/s = 7.0×10^{-6} M/s

Now they agree in terms of amount and dimension!!

Problem 1.3

Using the data of Table 1.8, calculate the average rate for the decomposition of N_2O_5, $-\Delta[N_2O_5]$; and the average rate of formation for O_2, $\Delta[O_2]$, in the time interval from t = 400 s to 700 s, if the reaction is as shows in Eq. 1.21.

Answer:

Average rate for decomposition of N_2O_5 = 6.66×10^{-6} M/s

Average rate of formation of O_2 = 7.0×10^{-6} M/s

1.8 Qualitative/Quantitative and Semi-Qualitative Analysis

If observable changes are of interest in a chemical process, which involves exclusively physical changes during a time frame (e.g., color, state, boiling), this analysis is called qualitative. However, when measurements take place during specific time intervals, it is defined as quantitative.

For example, the chemical reaction of equimolar quantities of pyrrole and aldehyde in propionic acid at 120°C leads to obtaining a porphyrin molecule by following the heterocyclic formation through UV-vis spectroscopy due to a gradual presence of the "Soret band" at 210–220 nm. However, the amount of porphyrin formed is negligible to analyze from the data collected every 20 or 30 min.

On the other hand, the detection of agricultural pesticides drained into fishponds is a topic for food and environmental concerns.

For this second analysis, not only detection but also quantification of pesticide concentration in soil/sediment must be considered to determine whether there are quantifiable pesticides in fish. If so, which ones?

Thus, spectroscopic protocols can be designed to address the concerns generated by both the determinations with the difference that in the first (porphyrin synthesis), a qualitative analysis is enough to screen for the presence of a characteristic functional group (heme), while in the second, not only the type but also the quantity of pesticide can be spectroscopically detected by not only addressing the presence of a hazardous substance, but also quantifying its concentration.

Despite quantitative analysis are determinant as R&D methods, another widely used analytical method on determine the presence of a specific substance where the concern is not exactly how much of this are there, but if it's presence in the sample is found in detectable quantities to be compared with a stablished (regulated) standard is the so called *semi-qualitative analysis*. In this category, the concern is not exactly how much substance is present in a sample, but whether the present substance is *above* or *below* a certain threshold level. For example, the *estimated average glucose* (EAG) is a standard used to find out sugar in blood. Thus, EAG is what we see when monitoring blood sugar at home using a sugar meter devise. Nonetheless, because we are more likely to check blood sugar in the morning and before meals, the personal meter readings will likely be lower than the patient's A1C quantified in a laboratory screening. Therefore, when it comes to a quantitative analysis, there is no one-size-fits-all target. Then, the A1C target levels can vary by a person's age and other factors, and the target may be different from someone else's. In that case, the goal for most adults with diabetes is to maintain their A1C levels less than 7%, which is equivalent to 140–154 mg/dl of sugar or 7.0 mmol/L of blood [8]. The goal here is to determine if the A1C of someone else is higher or lower than that accepted as the standard value, so if these levels are less than 6.5%, they are considered in the prediabetes range. Otherwise, if the analysis has an A1C level of 6.5% or higher, these values are in the diabetes range classification. With the last example, by performing semi-qualitative determination once the goal of analysis and the target analyte have been identified, the method available to carry out this type of analysis has to be reviewed periodically with extreme care to verify its accuracy, precision, cost, and other relevant constraints. Additionally, the amount of labor for sample preparation, time required to perform the analysis, and degree of automation can also be important tasks of consideration.

Chapter 2

Traditional Methods for Extraction and Separation of Natural Products

2.1 Introduction

Traditionally, plants (specially herbs) have been used as food for humans and domesticated animals, cosmetics, and fragrances. Plants are also known as the first natural source of traditional medicines in the treatment of a wide scope of illnesses and chronic diseases [9]. For a long time, traditional medicine has been defined by the World Health Organization (WHO) as "health practices, approaches, knowledge and beliefs incorporating plant, animal and mineral-based medicines, spiritual therapies, manual techniques and exercises, applied singularly or in combination, to treat, diagnose, and prevent illnesses or maintain well-being" [10]. Plants-derived products have gained increased importance worldwide for the treatment of many health-related problems. In this regard, the WHO has reported that about 5.6 billion people, which comprise ~80% of the world's population, use plants-derived products as their primary health care [11]. In this context, several plants-derived products have been processed as dietary supplements, nutraceuticals, organic health products, and traditional medicines. Thousands of plants-derived products are indexed in the Chinese and Indian national pharmacopeias [12].

Semi-Critical Assisted Extraction: Applications and Commercialization in Biotechnology, Food, and Pharmacy
Tulio Chavez-Gil

ISBN 978-981-4968-36-2 (Hardcover), 978-1-003-29124-4 (eBook)
www.jennystanford.com

According to the Reports and Data (2019), the global dietary supplements market was valued at $140.1 billion in 2018, with an expected increase to $216.3 billion by 2026, at a CAGR (compound annual growth rate) of 5.5%. Dietary supplements are industrialized plants-derived products, intended to supplement the diet that contains one or more of the following natural dietary ingredients: vitamins, minerals, proteins, and amino acids that are found in herbs, or other botanical ingredients. These nutritional supplements can be found in various forms such as pills, capsules, tablets, powder, or liquids [13]. An emerging trend in the industry of botanical ingredients is related with the identification of the different phytochemical families, each one of them responsible in part for the activity of the plant and the extract together, thus obtaining a phytochemical profile closer to that of the plant raw material, but radically different to that encountered in the conventional market extracts. The trend is enabling many companies to develop potent products that will help consumers gain health benefits from different organic sources [13]. According to the current data, more than 42% of the US population have claimed to have used dietary supplements of a botanical origin either to widen their current diet or to treat or prevent health-related issues [12]. Because the increasing costs of healthcare, people are turning toward dietary supplements to help them stay healthy. These dietary supplements provide various health benefits such as strengthening an individual's immune system, seasonal protection from cold and flu, prevention of migraine and headaches, treatment of arthritis, rheumatic diseases, chronic allergies, lower cholesterol, triglyceride levels and help regulate blood pressure, cardiovascular diseases, and cognitive diseases, all of them at lower costs.

2.2 Chemical Extraction of Plant's Active Compounds through Traditional Methods

2.2.1 Parts of Plant's Preparation

The extracted compounds of medicinal plants have often been applied for a broad range of acute and chronic conditions because they provide ample opportunities to gain a vast wealth of knowledge through the exploration of plants-derived products [14]. The phytochemical constituents extracted from medicinal plants such as anthocyanins, flavonoids, phenolics, and saponins are believed to have numerous therapeutic benefits that can reduce the risk of multiple diseases, including inflammatory conditions, cellular oxygenation, antibiotics, and cancer [15, 16]. Also, the chemical components present in plants contribute to their own protection against microbial and insect attack [17]. The extraction of folk medicines, such as flavonoids, has been utilized to improve human health condition via their multiple biological functions, including anti-inflammatory [18], antimicrobial [19], antioxidant [20], and anticancer activities [21], and the prevention of osteoporosis [22]. In this regard, at the beginning of the 19th century, the pioneering chemical analyses performed by scientists were used to determine the active ingredients from plant extracts, which subsequently led to the development of traditional medicines that had been passed historically through generations by word or practitioners and teaching. Thus, several traditional methods for chemical processing of plants have been performed with a common point between them and characterized to be the boiling of either the entire plant or selected parts of it and carried out by medical practitioners for the past 5000 years [23]. Nonetheless, the most empiric chemical extraction or separation procedure starts with the collecting process of plant raw materials, which may be submitted to extraction. To proceed, nonetheless, it is necessary to carefully select the plant parts that will be processed for physical separation and sample preparation. As indicated in Table 2.1, a general guideline for harvesting the three primary parts of a plant (i.e., leaves, stems, and roots), which enables its selection to provide the optimal results in a chemical extraction procedure, is necessary to follow. For example, it has been found that the *leaves* of *Labisia pumila* exhibited higher antioxidant activities and total saponin amounts than that in the plant roots and stems [24, 25]. The analysis of the extracted compounds has allowed the classification of four different stages of herbal processing to produce the final herbal active components [18]. Nonetheless, before an extraction process begins, the harvested plant parts must be pre-processed. In this stage, the parts must be dried to remove moisture for preservation, to prevent bacterial activity, and to restrain fungal growth [26].

Therefore, to improve the product yield, the dried plant must be ground first in a mortar or grinding system to increase this material surface area. This approach helps to improve the contact solute/solvent which works in favor of the subsequent extraction or separation stages. Increasing the surface area of dried materials will improve the performance of the extraction process, which will be discussed in advanced chapters. Another important parameter to take into consideration in chemical extraction technology is sample particle size, which is a key part of data used for statistical analysis to gather information on yield, time spent, solvent volume, or any other quantifiable information.

Table 2.1 Guidelines on harvesting plant parts for bioproduct extraction

Choice of Part to be Collected	Criteria for Best Harvesting Stage
Whole plant or aerial parts	When the plant is in its initial flowering
Leaves	Fully developed leaves but before their color changes
Flowers	At the beginning of spring or during the season where weather dominates flowers blossoming
Fruit	Fully matured if its juice is the target
Seeds	Removed from the fruit, dried, and grinded
Roots and rootstocks	After a plant achieves full maturity and stages of all parts are first determined
Stems or bark	At the beginning of the rainy season (Asia) when the maturity stage of the plant is first determined, and at the beginning of spring or at the end of summer (West)
Wood	Multiple sources can be harvested at different times of the year and mixed [27]

2.3 Marinated Extraction: Maceration

Marinated extraction or maceration (MA) is considered the simplest and, therefore, the most ancient extraction method and can be conducted at room temperature. The method has a great advantage when compared with advanced technologies for the extraction of thermolabile compounds. In this method, non-thermophysical parameters are considered the driving forces that account for it and excludes temperature, density, and pressure, but is highly dependent on the solvent's polarity (polar or nonpolar), as well on the ratio established between the sample and the solvent (m/v, v/v, or m/m) among of time spent. Once the extraction is considered complete, the obtained products are first separated through filtration with the filtrate rich in the desired solute(s). Another advantage of this empirical method is the fact that it does not require a special location such as a laboratory or any special equipment; therefore, simple glassware is sufficient, as shown in Fig. 2.1. In this case, a common glassware flask is used to conduct an inexpensive procedure without constant supervision that can be afforded after long periods, as for example, the extraction of total phenols from a plant bark of leaves or total anthocyanins from chokeberry fruit in a 50% ethanol solution under standard conditions. By using a solute: solvent ratio of 1:20 (m/m) and particle size of about 0.75 mm, maceration extraction is a simple (empirical) but very effective method for the extraction of phenolic compounds from different berry fruits [28].

Figure 2.1 Glass flask for marinated extraction.

Recent studies on the extraction of catechin (a natural compound with powerful antioxidant activity) from *Arbutus unedo* L. fruits were carried out using MA, microwave-assisted extraction (MWE), and ultrasound extraction (USE) techniques. The results showed that MWE is the most effective extraction method from the three. However, when temperature is decreased from that used for MWE, it was found that maceration extraction rendered similar yields of product as achieved through MWE, which can be translated into economic benefits [29]. An evaluation of the efficiency of polyphenol extraction from *Serpylli* herb was performed on various extraction techniques [e.g., MA, heat-assisted extraction (HWE), and USE]. Based on the total content of polyphenols, the USE method offered the highest yield of flavonoids, but not statistically significant difference was found between MA and HWE methods [30]. Chinese folk medicine uses *Cajanus cajan* leaves maceration extractions for the treatment of hepatitis, chickenpox, and diabetes, because this plant contains a large number of flavonoids as bioactive medicines [31]. Thus, a comparison of the extraction rates of orientoside (1), luteolin (2), and total flavonoids (3) from *Cajanus cajan* leaves was conducted by employing MWE, reflux extraction (RE), USE, and MA methods (structures 1, 2, 3). In the study was determinate that extraction efficiency for orientoside, luteolin, and total flavonoids has its lowest yield through maceration method [32]. Nonetheless, an interpretation to these findings lays with the fact that those polycyclic and poly-oxygenated molecules require more thermophysical intervention than simply room temperature and atmospheric pressure among polarity or density properties exhibited by a solvent under maceration extraction.

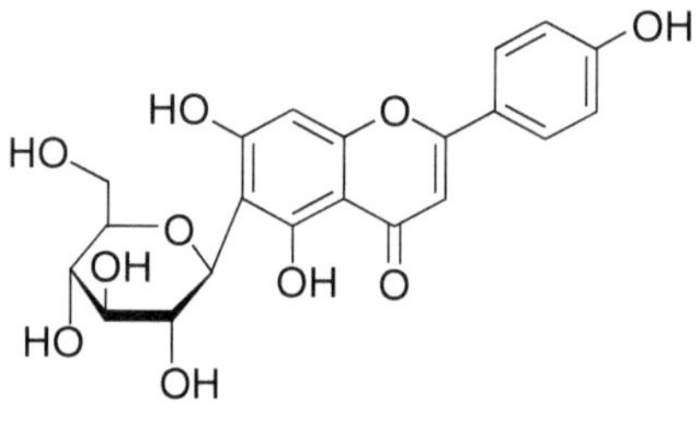

1-Orientoside

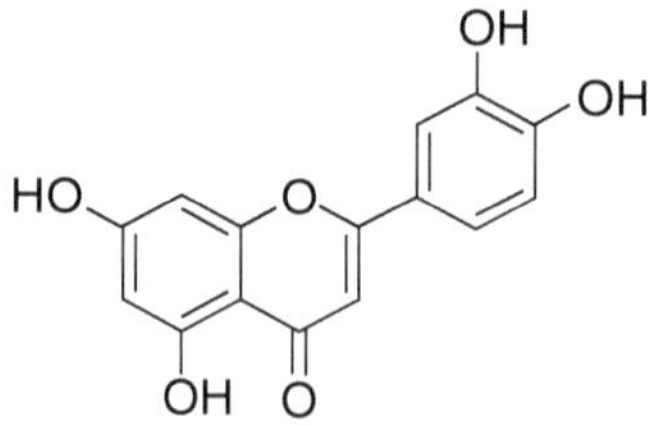

2-Luteolin

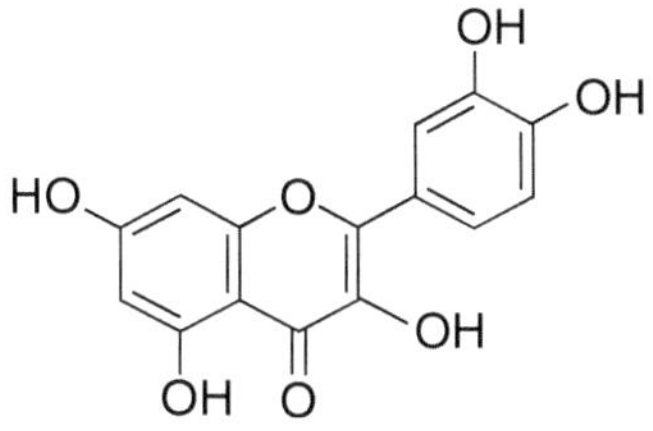

3-Flavonol

The most arguable concern regarding maceration extraction is the technological disadvantage associated with the long times required because the method does not use heat or other physical forces that act on the sample structure [28]. Nonetheless, the method, as described above, is very useful for both traditional medicine and cosmetology for the extraction of active compounds with extraction occurring within periods of 48 h at room temperature [33–35]. Although chemical mechanisms that explain the extraction of molecules from natural matrices are not yet available for the method, it is plausible to hypothesize that the process is dominated by the foundational rules of solubility [36, 37].

Thus, maceration extraction is a technique that applies for a compound in which its mass transfer to a solvent will depend on its capacity to solubilize in the extracted solution among its volatility (fugacity) equilibrium between the air and the solvent. In that case, a physical parameter that can measure the solute's volatility rate is needed to avoid trial and error during the process.

Because maceration is a procedure carried out at standard conditions, until certain extent, the process prevents volatilization or escaping tendency (fugacity) of a solute A from the extracted solution where A is partially or totally dissolved. The fugacity tendency, therefore, can be estimated if the physical equilibrium between the solute's gaseous phase, A_G, and its liquid phase, A_L, is determined from its distribution ratio K_D, also called Henry's law constant, H:

$$H = K_D = \frac{[A]_G}{[A]_L} \tag{2.1}$$

The constant as proposed herein, is a dimensionless physical quantity with the only purpose to explain that during a marinated (= maceration) extraction, a physical equilibrium is achieved when the concentration of A (gas or liquid) is equal in both phases (vapor/liquid). Nonetheless, the efficiency of the method can be determined by the fact that the larger the magnitude of Henry's law constant ($A_L/A_G \leqq H \geqq A_G/A_L$), the greater the tendency of A for its volatilization from the liquid solvent into the gaseous phase. Therefore, when physical equilibrium is achieved, Henry's law constant H can be estimated by measuring the concentration of A in both the gaseous and the liquid phases. In practice, however, the concentration of a semi-volatile to volatile solute is more often measured in one phase while its concentration in the second phase is determined by mass balance [38].

Alternatively, for a dilute to neutral or nonvolatile compound, Henry's law constant can be also estimated from the ratio of the solute's partial vapor pressure, P_{vp}, and its solubility, S, by taking the solution's molecular weight into consideration and expressing S as the solute's molar concentration:

$$H = \frac{P_{vp}}{S} \tag{2.2}$$

where P_{vp} is given in atm and S in mol · m^{-3}; therefore, H is given in units of atm · m^3 · mol^{-1}.

2.3.1 Solubility Effects on Maceration Extraction

In general terms, solubility is defined as the maximum amount of a chemical compound that can dissolve into a solvent in a fixed amount and at a given temperature (discussed in chapter three), but this chemical property can also be experimentally determined or physically estimated from Henry's law constant H, directly from the solute molecular structure, by avoiding a separate determination of vapor pressure and solubility, respectively [39]. In that case, for a solute present in a dilute concentration, the phase changes are very thermophilic and directly dependent on the temperature parameters, which can be expressed as:

$$\frac{H_1}{H_2} = \exp\left[\frac{\Delta H_S}{R} \cdot \left(\frac{1}{T_1} - \frac{1}{T_2}\right)\right] \tag{2.3}$$

with subscripts 1 and 2 in Eq. 2.3 referring to the two temperatures where the solute changes its state of phase.

Nonetheless, Henry's law constant H is very temperature sensitive; thus, a temperature correction may be necessary to estimate the solute's enthalpy change (ΔH_S J/mol) from a gaseous to liquid phase [39].

In addition to that, any other factor that can influence the activity coefficient of a solute dissolved in water may influence its solubility and, therefore, may affect the H values. In consequence, it will be notable to highlight herein that the presence of electrolytes, or another dissolved organic chemical as well as the surface of an active material, a colloid substance, or suspended matter, will also impact the chemical stability of the solute with direct effect on the distribution coefficient value too.

However, in the marinated method, one can assume that the solubility exhibited for the solute (as a semi-volatile chemical), which is extracted at room temperature and one atmospheric pressure, the extraction obeys Henry's law constant. Therefore, Henry's law can be used to analyze the **solute ↔ solvent** chemical equilibrium for a semi-volatile compound dissolved in either a polar or a nonpolar solvent if it is found in low concentration. Because the marinated extraction is performed on materials characterized for presenting low volatility, it is assumed that the solute in the extract solution is a semi-volatile molecule in a diluted concentration. Therefore, Henry's law constant, H, can be calculated from the ratio of the solute's vapor pressure, P_{vp}, and its solubility, S, from Eq. 2.2, which can be converted to the dimensionless Henry's law constant, H, by assuming its dependency of temperature as:

$$H = \frac{P_{\text{vp}} + (WM)}{0.062 \cdot S \cdot T} \tag{2.4}$$

Commonly, the concentration of solute in the extracted solution after maceration extraction is found mainly in diluted quantities; the partial vapor pressure (P_{vp}) of this type of solute is given in mmHg, rather than atmospheres; the molar mass (MW) of the extract is the solute's molecular weight; S is the solute's water solubility constant in mg/L, rather than mol/L; T is the temperature in Kelvin. And for diluted solutions, the quantity 0.062 is considered the most appropriate for the universal gas constant that applies for an aqueous low concentration solution [40].

Nonetheless, the escaping (or evaporation) tendency of a solute from an organic liquid solvent to the gas phase had been usually classified as high, medium, or low [41] by taking into consideration two physical properties responsible for the tendency such as vapor pressure P_{vp} and solubility S:

1. Vapor pressure of solute in an organic solvent (pressure dependent)
   Low: $1 \cdot 10^{-6}$ mmHg
   Medium: between $1 \cdot 10^{-6}$ and $1 \cdot 10^{-2}$ mmHg
   High: greater than $1 \cdot 10^{-2}$ mmHg
2. Solute water solubility as parts per million ($1 \cdot 10^{-6}$ parts)
   Low: less than 10^1 ppm
   Medium: between 10^1 and 10^3 ppm
   High: greater than 10^3 ppm

In the last approach, however, the concentration is expressed in parts per million (ppm), a unit which does not define the solute's molar mass nor its molecular structure [42]. Thus, this unit (ppm) does not consider the chemical identity or the molecular character of the solute.

Therefore, a rearrangement of Eq. 2.2 produces:

$$P_{vp} = \mathrm{H} \cdot \mathrm{S} \qquad (2.5)$$

The linear equation 2.5 is a solution to Eq. 2.2, and a plot of vapor pressure (y-axis) versus solubility (x-axis) yields a slope (see Fig. 2.2) that represents Henry's law constant with different values of H by considering the above three possibilities regarding the two factors vapor pressure and solubility applied for an organic or water solvent calculated at low concentration and constant temperature.

The slope values calculated from these two variables can be interpreted as follows: A solute (a solid product dissolved in a liquid) presenting low volatility in the liquid solution is characteristic of organic molecules that have low vapor pressure and high solubility range, whereas the highest volatility values of a liquid solution are exhibited by molecules with high vapor pressure and low solubility, as occurs, for example, on hydrocarbons having less than five carbons by molecule.

Similarly, intermediate levels of volatility must result from the vapor pressure and solubility combinations taken together.

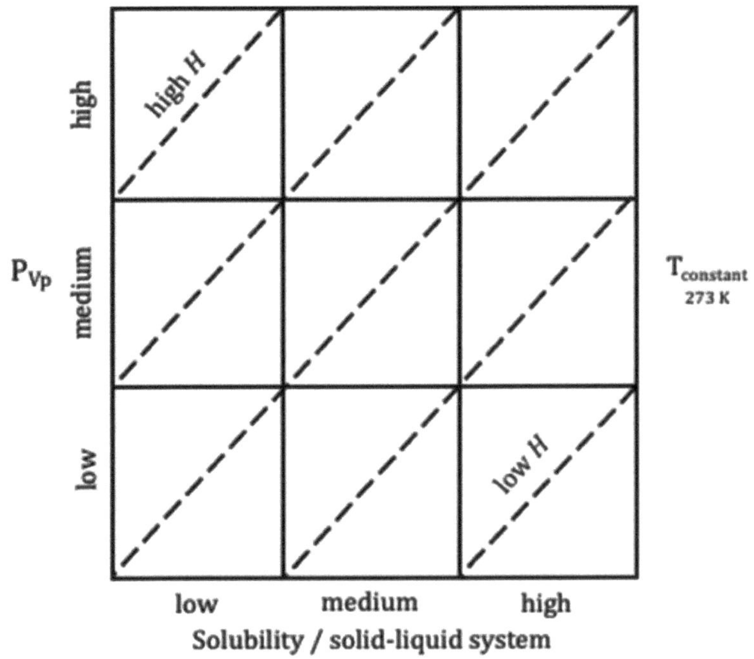

Figure 2.2 Diagram depicting slopes as Henry's law constant *H* and conceptually represented by diagonal (dashed) lines as plot of vapor pressure P_{vp} versus solubility *S* for a low concentration solid–liquid system at constant temperature.

So the quantitation of *H* can be determined from the *solute–solvent* ratio in mixtures with low vapor pressure and low solubility, medium vapor pressure and medium solubility, or dominated for high vapor pressure and high solubility, which exhibit nearly equivalent volatility from an organic liquid solution.

Thus, in a practical analysis of this type, Henry's law constant can be used as an analytical tool regarding the choice for the most affordable extraction technique according to the solute's volatility in the extracted solution attained.

For example, if the value of Henry's law constant for the analyte is less than Henry's law constant for the solvent alone, then the solute is nonvolatile (i.e., 99% is dissolved) in that solvent and, therefore, the solute concentration will eventually increase as the solvent evaporates at standard conditions.

On the other hand, if Henry's law constant for the analyte is greater than Henry's law constant for the solvent, then the solute is classified from semi-volatile to volatile in the solvent, including at standard conditions.

In this case, if the late solution is open to the atmosphere, the volatile solute's concentration must decrease in the time scale, because the solute may evaporate more rapidly than the solvent due to its high vapor pressure characteristics which obeys more those of the gas phase, and therefore, the H values has inverse proportions which lays with adverse chemistry consequences regarding both the efficiency of the extraction and a decrease on the recovered products' yield.

By considering that the most common gas–liquid equilibrium encountered in chemical extraction is exhibited by the air–water interface, it is not coincidental that the *maceration method* has been used for a long time due to the advantages offered by the solubility property that some organic molecules has in water (e.g., proteins, vitamins, minerals) and that according with this brief physical analysis, the best results are obtained when the extract is rich in semi-volatile to nonvolatile material that doesn't susceptible to high temperature and that not apply for the method in discussion because the common solvent used in it is water.

As a practical argument, the application of Henry's law constant for this traditional extraction method leads to the hypothesis that the volatility presented for any solute in marinated extraction can be used to determine the solute's solubility properties to offer the possibility of quantifying its vapor pressure and solubility constant, which provides important guidelines for organic solutes when they are dissolved in a solvent other than water by considering the data in Table 2.2. In addition, it will be useful to keep in mind if a solvent belongs to one of the following categories before its use in an extraction process:

- **Nonvolatile**: Volatilization is unimportant for $H < 3 \cdot 10^{-7}$ atm $m^3 \cdot mol^{-1}$ (H for water itself at 15°C).
- **Semi-volatile**: Volatilizes slowly for $3 \cdot 10^{-7} < H < 10^{-5}$ atm $\cdot$ $m^3 \cdot mol^{-1}$.
- **Volatile**: Volatilization is significant in the range $10^{-5} < H < 10^{-3}$ atm $\cdot$ $m^3 \cdot mol^{-1}$.
- **Highly volatile**: Volatilization is rapid if $H > 10^{-3}$ atm $\cdot$ $m^3 \cdot mol^{-1}$.

Table 2.2 Dimensionless Henry's law constants, H_C, with solubilities and vapor pressure of some chlorinated and non-chlorinated compounds at constant temperature

Solvent	H_C^1	S^2	S^3	P_{vap}^4
Benzene	0.22	22.8	—	0.123
Ethylbenzene	0.32	1.4	—	0.012
Toluene	0.27	5.5	—	0.125
o-Xylene	0.20	1.6	—	0.008
m-Xylene	0.28	1.5	—	0.010
p-Xylene	0.29	1.7	—	0.011
Tetrachloroethylene	0.72	0.8	5	0.024
Trichloroethylene	0.39	8.4	8	0.097
1,1-Dichloroethylene	1.07	4.2	25	0.298
Cis-1,2-Dichloroethylene	0.17	36.1	35	0.267
Trans-1,2-Dichloroethylene	0.38	65.0	31	0.435
Chloroethylene	1.14	43.2	13	5.360
Ethylene	8.75	0.8	11	68.55
Tetrachloromethane	1.24	7.5	—	0.220
Trichloromethane	0.15	66.2	—	0.250
Dichloromethane	0.09	228	—	0.572
Chloromethane	0.36	106	—	5.657
Methane	27.2	1.5	—	1.97*
1,1,2,2-Tetrachloroethane	0.019	17.9	—	0.007
1,1,1-Trichloroethane	0.70	5.4	—	0.031
1,1,2-Trichloroethane	0.048	33.1	—	0.162
1,1-dichloroethane	0.23	51.5	—	0.298
1,2-Dichloroethane	0.044	87.9	—	0.103
Chloroethane	0.46	88.0	—	1.328
Ethane	20.4	2.0	—	1.97**

[1]Dimensionless Henry's constant at 25 °C (Gosset 1987: Mackay and Shiu 1981).
[2]Aqueous slolubility in mM at 25 °C (Mackay and Shiu 1981).
[3]Aqueous solubility at 23 °C (Fox et al. 1990).
[4]In atm at 25 °C; *–152.6 °C; **–75 °C. Adapted and modified from Ref. [44], and PubChem
www.pubchem.ncbi.nlm.nih.gov, and https://www.engineeringtoolbox.com/ethane-d1417.html

From the above discussion, this empirical extraction method can be applied as common and inexpensive technique to extract low volatile chemicals from plants origin if a careful setup is followed regarding prevention of solutes to volatilize throughout the extraction mechanisms by make a careful selection of the solvent where the extracted molecules achieve both high solubility and low vapor pressure once the process stop.

2.4 Percolation

Percolation is also classified as an empirical method, credited to be more extraction efficient than maceration because the process is carried out on continuous loading of solvent through the sample. Similar as occur to maceration, the amount of solvent could be denoted as saturated because its amount is several times greater whether compared with the sample [43].

In addition, the sample is kept into a stationary container (mainly a filter) while a continuous replacement of hot fresh solvent, pass throughout the sample such as occurs in a homemade coffee maker (Fig. 2.3).

Figure 2.3 Typical percolation extraction system (a household coffeemaker).

In simple words, the procedure is carried out with the sample at room temperature, which literally means "pass a liquid through a solid material drop by drop" with the solvent slowly passing through the plant part (e.g., leaves, seeds, flowers) little by little, and loading itself with the chemical active ingredients dissolved in the solution, which is slowly pushed downward by the solute by the pure solvent constantly added from above.

2.4.1 Sample Preparation and Procedure Description

Before the plant part or selected sample of interest is placed into the percolator (a filter), it must be appropriately shredded by being careful in not making its particles too fine to avoid them to load together within the filtrate solution. If overloading occurs, this could cause difficulties in separating the solid sample from the extraction liquid (the solvent), causing the extract to be cloudy and contaminated with the solid residue, which is detected by the naked eye in the receiver's bottom.

To avoid this possibility, it is advised to preventively moisten or dampen the plant raw material with a small volume of the chosen solvent for the extraction as well as making this moistening procedure with the filter to prevent the solvent from overflowing the filter by carrying and downloading the unwet sample.

By following this preventive step, it helps to initially expand the cell wall of the plant's part for better extraction of its chemicals, allowing an easy passage of solvent through its matrices, which will accelerate the solvation of active ingredients that are finally dissolved in the liquid solvent.

Once the filter and sample are preventively wet, the solvent is fed from above the sample, which passes through the material at a rate that is determined by the volume added, and until certain point, this rate is dependent on the solvent temperature to undergo extraction. The feeding speed must not be excessive; otherwise, the solvent could not have time to penetrate the matrices' cell and to excerpt the extraction of molecules contained within the matrices; and in the opposite way, if the solvent's flow is too slow, its temperature could decrease and probably more solvent should be needed to complete the extraction. A recommended speed for solvent flow supply for about 1 kg of sample must not be greater than 5 ml/min. Based on the chemical nature of the active compound(s) to be extracted (e.g., oil, ionic, hydrocarbon, fatty acid, oligosaccharide, etc.), the choice of solvent varies, but in general, a hydro-alcoholic solvent is used, which means the solvent is made up of a mix of water and potable alcohol.

The mix will produce a more efficient solvent for extraction because water keeps the cell wall hydrated, while the nonpolar side of alcohol, which is chemically similar in nature to most of the active ingredients, acts on the less polar molecule(s) by leading them to dissolve. The use of alcohol for extraction is considered pharmacologically advantageous in terms of its capacity to preserve most molecules free of oxidation for a long time [34]. An indication that percolation was complete relays with the fact that the liquid that exits the percolator is colorless and, therefore, devoid of sample active ingredients.

The first fraction obtained is defined as the head extraction and contains the highest number of extracted substance(s). The fraction obtained, rich in active ingredients, is called "leachate," and once the percolation has ended, the sample used for extraction, still soaked in the solvent, undergoes some pressing to enhance recovering of residual liquid absorbed by cell matrices, and this liquid is then added to the leachate. The method can be used for successive partial leachates by employing a different solvent to carry out each extraction after perceiving that the amounts of the active ingredients have diminished in extract. However, percolation is an extraction technique not recommended in the case of plants that tend to expand a lot (e.g., plants containing pectin), which causes swelling and high solvent retention. Therefore, the solvent cannot flow correctly as pectin molecules have the tendency to form gels that clog pipes [44, 45].

2.4.2 Modified Percolation

The conventional method of percolation can be modified by including evaporation of the more concentrated product when the extraction is performed by employing as solvent-diluted alcohol. In the process, the strength of the alcohol needs to be unaffected by the concentration of extract(s); therefore, the percolation is applied as a continuous process and the first fraction of percolate is collected and set aside. The subsequent fractions of percolates are collected and concentrated, and the first fraction of the percolate is then added to the final product. With this sequence of steps, the alcohol strength required to maintain an optimal extraction performance produces the higher concentration of products. This process is known as the reserve percolate method and has been developed in instruments constructed by researchers, as shown in Fig. 2.4. In these modified processes, the percolation technique acquires more technological character as continuous or semi-continuous extraction, which are processed in devices that are designed as batches possessing varying sizes [46–49].

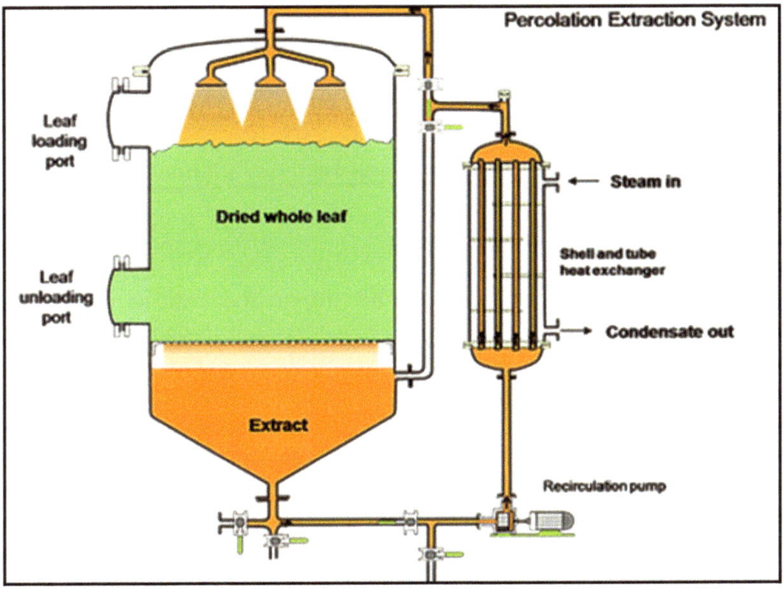

Figure 2.4 Diagram of fixed bed reactor for continuous percolation extraction. Reprinted from Ref. [50].

This extraction battery consists of several vessels designed in series, which are interconnected through feeding pipelines and are arranged in such a way that the solvent can be added through several feeding ports. At the end of the process, the product is removed from any vessel and such type of extraction battery gives maximum efficiency of extraction with minimum use of solvent. The product obtained in the process is more concentrated, and less loss of solvent takes place due to low evaporation. However, the analysis of yields for the analytes collected with this modified method shows that the process efficiency is highly dependent on the response of the plant parts' structure to solvent chemical activity. Therefore, these are some advantages and disadvantages encountered in scaling up this modified method:

Advantages:

- The method ensures less liquid material handling and processing.
- If lower water is removed from the extract, it can be purified by doing the process more efficient in energy use.
- The system can be thermally insulated by ensuring high thermal efficiency.
- The process is set up to comply with more hygiene regulations.
- The product recovered is relatively free of suspended material because filtration can be improved using membranes instead of cellulose as filter.

Disadvantages:

- Batch extraction cannot be integrated into a continuous process system.
- Repeated extractions are necessary to achieve good efficiencies.
- After a certain operation time, leaves become soggy. Then the hydraulic conductivity of the leaves in the bed decreases, which impairs percolation, thus increasing extraction time.
- Leaf loading and unloading through hatches is time consuming and results into longer time between the batches.

2.5 Water Extraction

2.5.1 Hot Water as Subcritical Solvent Extraction

Hot water extraction (HWE) belongs to the same category as the accelerated solvent extraction (ASE) and the subcritical water extraction (SWE) methods. The HWE and SWE methods utilize hot water at different temperatures (100 to 374°C) and pressure high enough to maintain the solvent in its liquid phase for extraction instead of organic solvents as is shown in Fig. 2.5 [50]. An advantage on the use of water as solvent is due to its lower cost for extraction operations if compared to another polar solvents (e.g., alcohols, acetone, halogenated, nitrogenated), otherwise if water does not need to be an ultrapure or bi-distilled solvent as required in some pharmaceutical or biotechnological extraction/separation processes. As a solvent, water is relatively easier to use, recover, and dispose and possesses relatively less issues related to environmental regulations. In addition, water possesses unique physical chemistry properties such as a disproportionately high boiling point if compared with its molar mass and high dielectric constant among its high polarity [50].

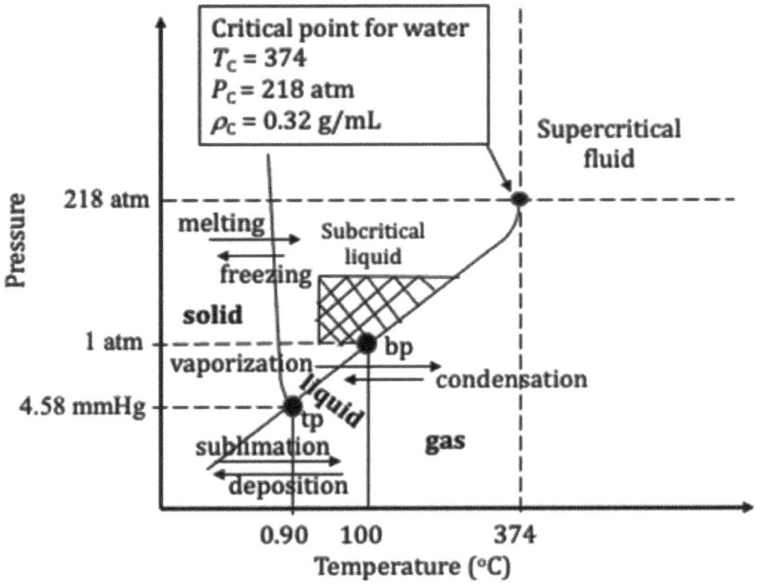

Figure 2.5 Phase diagram of water as a function of temperature and pressure with cross-hatched area indicating the preferred region for all forms of hot water extraction (HWE) and subcritical water extraction (SWE). Adapted from Ref. [46].

In the hot water chemical extraction, as the temperature rises, there is a marked and systematic decrease in permittivity, an increase in the diffusion rate, and a decrease in the viscosity and surface tension [50].

In consequence, more polar materials with high solubility in polar solvents (water like) at room temperature and pressure are extracted most efficiently at lower temperatures, whereas moderately polar and nonpolar molecules require a less polar medium induced by elevated temperature.

The extraction of oregano essential oil through hot water extraction has been shown to be faster, cheaper, and reliable in comparison to hydro-distillation extraction [51]. In terms of efficiency, it has also been found that the hot water extraction method can extract the most valuable antioxidants (e.g., carvacrol and thymol) from the essential oils of *Thymbra spicata* at high concentration (~90%) than using organic solvents [52]. Table 2.3 shows experimental results on extraction time relative to square deviation analysis, SD percent, retention indices (RI) on the DB-5 column among the amount of thymol and carvacrol (milligram per gram dried sample) performed through hot water extraction, hydro-distillation, and Soxhlet extraction, respectively [38].

Table 2.3 Thymol and carvacrol (mg/g dried sample) amount of essential oil of *Z. multiflora*, extracted by hot water (SWE), hydro-distillation, and Soxhlet extraction methods

Components	SWE*	Hydro-distillation†	Soxhlet‡	RI§
Thymol	9.25 (4.77%)**	4.38 (2.97%)	0.94 (2.78%)	1,232
Carvacrol	11.51 (4.33%)	4.06 (3.31%)	1.39 (2.83%)	1,242

*Extraction time = 150 min
**Relative SD percent
†Extraction time = 180 min
‡Extraction time = 210 min
§etention indices (RI) on the DB-5 column
Reproduced from Ref. [46]

The data show that the amount of valuable oxygenated components extracted under the SWE method is significantly higher than those extracted through hydro-distillation and Soxhlet extraction, respectively. In a comparative extraction, hexane was employed as nonpolar solvent and consequently, non-oxygenated components in the extract were enhanced whether compared to those extracted by the hot water method.

On the other hand, non-oxygenated components presented lower vapor pressures when compared to that of oxygenated compounds, and therefore, the oxygen content in hydro-distillated extracts was visibly increased compared with the Soxhlet extraction ones. Because of the significant presence of the oxygenated components, the final extract using hot water extraction methods was relatively better and more valuable. The importance of carvacrol and thymol compounds lies in their high content of valuable nutrients for the food and nutraceutical industry, and a comparison of the extraction efficiency between SWE, hydro-distillation, and Soxhlet methods has been done on the extraction of essential oils from coriander seeds (*Coriandrum sativum* L.) [53].

When the three extraction technologies are compared by using water as solvent, the findings showed that hydro-distillation and Soxhlet extraction presented the higher extraction efficiencies than hot water extraction for low-oxygen compounds. However, hot water extraction offers better results in the extraction of more concentrated and valuable oxygenated components in essential oils widely used in food industry bioproducts.

Extraction of essential oil from marjoram leaves was done to compare the performances of hot water and hydro-distillation extraction methods. It was found that the volume of essential oil extracted using SWE was 5.1 fold more than the volume extracted through the hydro-distillation method [54]. Other authors have also reported that the extracted essential oil by SWE contained a high number of oxygenated components than those extracted with other methods [55]. Therefore, besides the efficacy of oil extraction and the selectivity of oxygenated compounds, the hot water extraction method can extract the same amount of essential oil within 15 min when compared to hydro-distillation, which took 3 h for the same experiment [56, 57].

From scheme A in Fig. 2.6, it shows that in hydro-distillation extraction, both the sample and solvent (water) are placed together in a retort or refluxing flask, and heat is applied to boil the solvent and vaporize the mixture with water steam to create small sacs (bubbles) in the plant raw material containing the essential oils. The exposition of plant cells to high vapor pressure makes the cells to swell by a large amount of vapor and condensed water; and eventually they burst, releasing their content (e.g., oils, volatile molecules, nonvolatile ones, hydrolysable ions, oxygenate compounds), and recover after the extraction is completed.

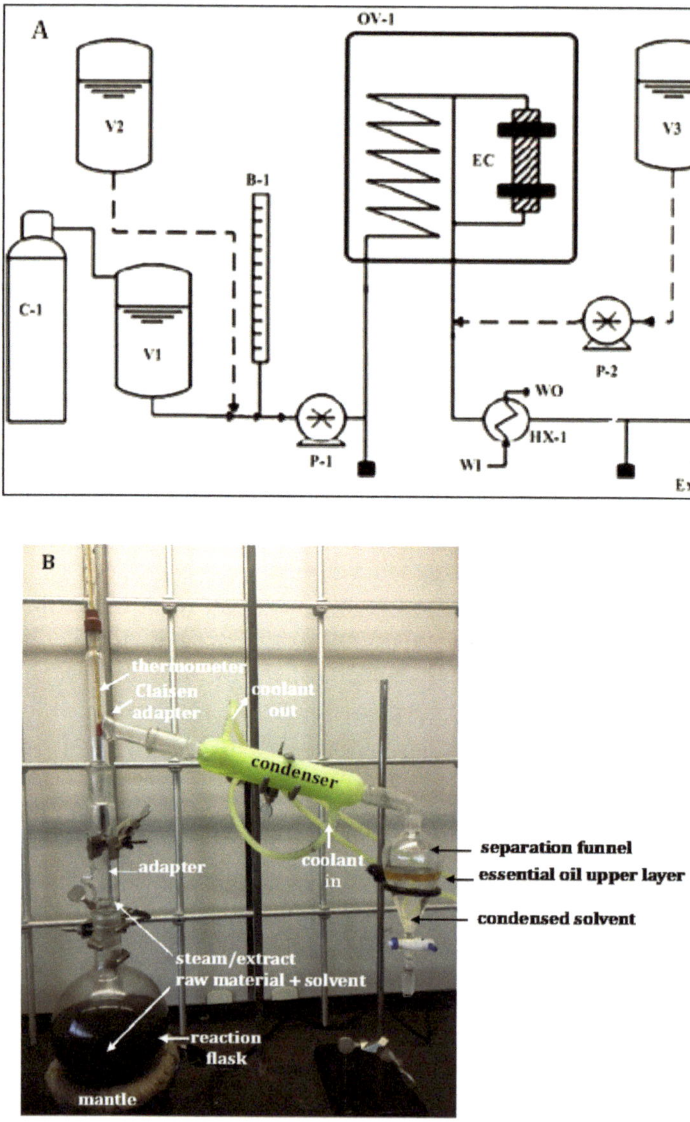

Figure 2.6 Schematic diagram of a hot water extraction system: (A) where B-1: Burette, C-1: Nitrogen cylinder, EC: Extraction cell, HX-1: Heat exchanger, OV-1: Oven, P-1, 2: Pumps, V-1: Water tank, V-2: Solvent tank, V-3: Rinsing solvent tank, WI: Water inlet, WO: Water outlet, and (B) glassware assembles of a hydro-distillation extraction system. Produced from Ref. [46] and picture from author's lab, respectively.

In hydro-distillation, the extracted materials are transported by steam in the vapor phase into the condenser where the condensate liquid mixture is formed. The liquid mixture then flows into a separator container where water and the essential oil are separated in two phases by density differences.

When the water-rich phase contains some plant essence that escapes to air at room temperature, it is called "hydrosol." For example, if roses are extracted using this method, the hydrosol is retained for use as a mild antiseptic and as fragrance for floral aromatherapy [42]. A claimed disadvantage of this method is its need for constant supervision and observing precautions with temperature limits inside the reaction flask to prevent thermolabile compounds degradation.

Thus, as a precaution, lavender essential oil extraction must be carried out at temperatures between 373 to 391 K (100–118°C). If, for example, the extraction of lavender essential oils occurs at temperatures above 391 K, it will result in oils containing more chemicals than the targeted aroma, thus causing the therapeutic effect of oils to be reduced or even neglected. In addition, hydro-distillation extraction is also recommended to be performed within a specific time frame to ensure that essential oils and other beneficial components are completely extracted. Although hydro-distillation extraction is a method that does not require expensive organic solvents, it involves sometimes several separation steps, necessary to separate liquid organic components from the liquid oxygen-rich extract dissolved in water.

The separation process could require, in some instances, long periods for settling out among the use of additional solvents that facilitate selective separation [43].

Chapter 3

Conventional and Semi-Automatic Extraction Methods

3.1 Introduction

This chapter focuses on two widely used techniques for the extraction of miscible/immiscible compounds naturally existing as liquid, gel, or solid material: (1) liquid–liquid extraction (LLE) and (2) solid–liquid extraction (SLE). Traditional and improved techniques have been useful in selected circumstances, but these two techniques have become the extraction methods of choice for research and analytical laboratories since a long time ago. In addition, a third semi-automated and widely used technique, *accelerated solvent extraction*, is also discussed. To understand the principles that govern either an empirical (traditional) or a sophisticated extraction technique, it is undoubtedly necessary to take into consideration some underlying physics and chemistry foundations surrounding the extraction procedure we selected. Hence, to succeed in a chemical extraction, a researcher or technician needs to have a clear knowledge of the chemical properties of the desired solute (the product) as an important factor to consider during the extraction–separation proceedings, among the inside–outside optimal physical conditions to set up (temperature, pressure, time) and necessary to achieve an efficient performance with the selected or in-place

Semi-Critical Assisted Extraction: Applications and Commercialization in Biotechnology, Food, and Pharmacy
Tulio Chavez-Gil

ISBN 978-981-4968-36-2 (Hardcover), 978-1-003-29124-4 (eBook)
www.jennystanford.com

extraction equipment. Among the aforementioned factors, it is highly recommended to be clear on the physical state of the solvent chosen to be used, such as if it is a gas, liquid, or supercritical fluid, to proceed under the rules of safety and to effectively extract or separate an analyte (solute) from its matrices.

Therefore, from the aforementioned relevant experimental considerations, the sample properties together with solute and solvent mass distribution, a selected group of physical chemistry properties are considered fundamental to understand the extraction process, including:

- The analyte's physical charge (polar, nonpolar) responsible for its solubility,
- Its molecular weight (if known),
- Its vapor pressure (if known),
- Its acid/base dissociation (its ionic character), and
- Its thermostability.

These physical chemistry properties are considered pharmacologically essential to determine the mass transport phenomena for the chemical constituents in the human body, as well as the mass transport of chemicals in the air–water–soil environmental compartments and are the foundations in our understanding of the transport phenomena between immiscible phases as taking place during an analytical extraction and its batch separation procedures [58].

3.2 Solubility: A Chemical Perspective

According to a compound's solubility of an either bioactive or raw material that is suspected can be extracted from a defined or diverse sources that comprises but are no restricted to plants (e.g., bark, seed, roots, leaf), fruit, crop, algae (macro-, micro-), or from animal origin (egg, milk, meat, protein, vitamin, immunoglobulin, fish) there are a large number of inorganic, organic, polar, nonpolar, among of mixtures of solvents to perform the extraction or separation by follows an even chemistry, biochemistry, biotechnological, or pharmaceutic method. If the substance of interest is lipophilic, for example, the solvent of choice will be organic nonpolar, ranging from those with a very low polarity (such as hexane) to those that are less nonpolar (such as chloroform and dichloromethane) (see Fig. 3.1).

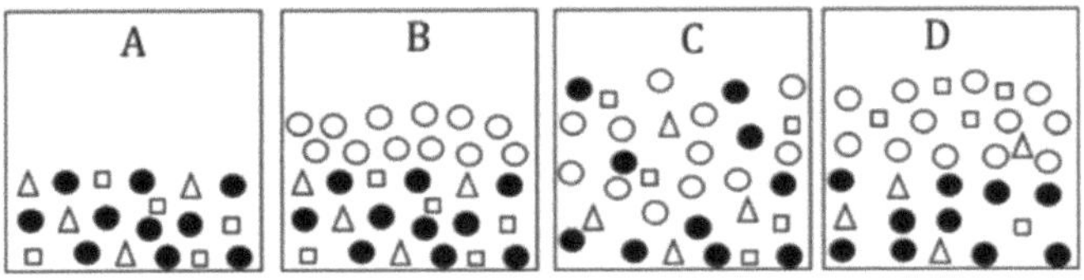

Figure 3.1 Two compounds (Δ, □) are placed on a solvent • (A), a second solvent O is added (B), but it is not soluble in solvent A; the solvents are mixed under stirring (C); and after the mixture is allowed to settle out, solvents separate again and one compound will go to one solvent or the other based on their polarity (D).

With this information in hands, the most commonly used nonpolar solvents for these types of extractions are cyclohexane, hexane, toluene, benzene, diethyl ether, and petroleum ether; with the extraction community considering these as the most effective solvents for the extraction of alkaloids, coumarins, fatty acid methyl esters (FAME), flavonoids, anthocyanins, and terpenoids [59].

On the other hand, for hydrophilic compounds, the choice will fall on the use of polar solvents, which include non-protic ones, such as acetone, or protic ones, like ethanol, methanol, or even water. In fact, acetone, acetonitrile, butanol, propanol[iso], and ethanol are the best choice of solvents for the extraction of flavonols, lectins, alkaloids, flavones, polyphenols, tannins, and saponins [60]. One of the major pros with solvent-liquid extraction lays in the use of simple equipment of low cost whether compared when gases are employed for same purposes.

In either solid–liquid or liquid–liquid extraction/separation, the process is based on the chemical principle that "a solute" can distribute itself in a certain ratio between two immiscible solvents, as depicted in Fig. 3.1, where one of them is usually polar (water) and the other is an organic (with low polarity to non-polarity) solvent such as acetone, acetonitrile, ethyl acetate, cyclohexane, n-hexane, benzene, dichloromethane, chloroform, diethyl ether, and linear or branched alcohols.

In certain cases, the solute can be more or less completely transferred into the organic phase; therefore, the technique can be used for processes such as preparation, purification, enrichment, and separation, that in all scales of working fronts leads from lab microanalysis to large-scale production.

These broad possibilities make solvent extraction a universal technique as the most affordable and widely used for instruction labs (academia), biotechnology as well as scaled industrial processes.

Current laboratory trends, however, are away from this technique, mainly because of the costs to adjust the process with states and international environmental regulations and because it has become more difficult and costly to dispose of the mixed waste generated from large volumes of solvents required for large-scale industrial extraction.

The technique also tends to be labor intensive due to some processes requiring multiple separation steps, which are carried out by using a battery of separatory funnels. Nonetheless, solvent extraction remains a powerful separation technique worthy of consideration for food, fragrances, pharmaceuticals, nutraceuticals, and biotechnology application.

By using a mix of solvents as the separation technique, it is explained as a partitioning process based on the unequal distribution of a solute (s) between one, two, or more immiscible solvents (S), with one of them usually being water (S_{aq}), or an ionic solution (S_{ion}), or an organic liquid (S_{org}):

$$s \rightleftarrows \mathrm{S}_{(\mathrm{aq,\ ion,\ org})} \tag{3.1}$$

If a chemical equilibrium is achieved in one or more steps during the process, the solute s can be found as solid, gel, or liquid, but in less concentration, or in different polarity whether compared with the solvent's amount and/or solvent polarity. As expressed above, the extracting solvent can be pure water, a water-miscible solution, or a water-immiscible solvent; nonetheless, the solute must be little to non-soluble in one of the solvents that forms the liquid mixture. To ensure a total separation, we need to understand that solutes exhibit different solubilities when exposed to various solvents; therefore, a careful choice of most appropriate extracting solvent will depend upon our knowledge of the physical chemistry properties exhibited by the solute, the solute/solvent phase density (or viscosity) as well as other parameters depending on the extraction method we choose.

In general terms, the solute can be removed along with its impurities, which are commonly called "interfering analytes" as the impurities can dissolve in one of the solvents (e.g., S_{org}), which is present in a large percentage in the mix.

It is important to highlight at this point that if an aqueous or polar organic solvent is added to the liquid extract and the two are thoroughly mixed, usually by shaking the extract mix in a separatory funnel, the shaking activity produces a fine dispersion of each solvent into the other, which will separate eventually into two different layers after standing from a few to several minutes after being exposed to standard room conditions.

On inspection with the naked eyes, it is not difficult to observe that the solvent possessing the highest density will form the bottom layer. In this stage also, a visual separation between the layers is achieved because the solute and its accompanying impurities (or interfering analytes), which have different solubilities in the solvents, can be separated and remain dissolved in each or only one layer. Once the two phases separate, the solute might preferentially remain dissolved in the organic phase (nonpolar), while the impurities from the solvent or the extract solution can selectively dissolve in the aqueous or the polar covalent phase too, but in low concentrations. In the opposite phase of solutes, the impurities and other interfering particles can be extracted from the pure organic or covalent polar layer into the most polar layer by employing several fractionated extractions.

To improve the separation of impurities, an alternative procedure consists of making the solute more soluble in a polar organic solvent, which also improves the extraction of nonpolar particles into the polar (aqueous) layer by leaving the solute behind in the organic layer but free of impurities.

Furthermore, solubility is a property that can also be explained by the rules of equilibrium, which can be used to describe how a substance with a defined chemical composition interacts with one or a mix of solvents, with the solvents being characterized by presenting different polarity between them.

Nonetheless, the use of two solvents is the best choice to conduct extraction or separation. For example, when a dissolved chemical component Y from a liquid phase A is accomplished by bringing the liquid solution of Y in contact with a second phase B, leading to the formation of two phases (A and B), which are not yet miscible together at room temperature and at one atmosphere of pressure with the phase B being a liquid, gas, or a supercritical fluid product (commonly a foam-like).

After being sure of the phase's physical state (liquid, gas, or supercritical fluid), an unequal percentage distribution of the extracted component Y between the immiscible phases must occur at the beginning. Nonetheless, after a short time of contact between these two phases, a balanced mass distribution of the extract (Y) is achieved in the initial phase, with the extracted substance being released and/or recovered from phase B for a subsequent extraction procedure (batch) or through an automatic instrumental analysis technique (e.g., GC, HPLC).

Moreover, to recognize the chemical distribution of components around the extraction theory, it may be sufficient if the experiment practitioner is familiar with the comprehensible equilibrium equation that describes the "extraction process" as a quasi-reversible to reversible mass distribution equilibrium:

$$\mathrm{Y_A} \rightleftarrows \mathrm{Y_B} \tag{3.2}$$

In this equation, once the distribution of Y (solute or extract) between A and B phases achieves its maximum equilibrium, a constant related with its mass distribution, K_D, for the system can be expressed mathematically and the equilibrium constant of this expression is referred to as the Nernst distribution law or partition constant [4], defined as:

$$K_\mathrm{D} = \frac{[Y]_\mathrm{B}}{[Y]_\mathrm{A}} = \frac{[Y]_\mathrm{org}}{[Y]_\mathrm{aq}} \tag{3.3}$$

where K_D is the chemical equilibrium constant, and [Y] comprises the concentration of the solute Y being tested, with "org" and "aq" referring to the organic and aqueous phases, respectively. However, the IUPAC further recommends the expression may describe the "partition ratio" in cases where transfer activity coefficients (heat or mass) can be determined, and that "distribution ratio" might be used to define the ratio of total analytical concentrations of solute [Y] between immiscible phases, regardless of its chemical form [61].

Also, once the solute had achieved its equilibrium distribution in the two phases, the mole distribution of the solute [Y] in Eq. 3.3 denotes a particular concentration pattern (moles/L) of Y in each phase at constant temperature and pressure (understood as the activity distribution of Y in two nonideal solutions). By convention, it is assumed that the concentration of an extract into phase B may appear in the numerator of Eq. 3.3, as if it is the product in a conventional chemical reaction.

At equilibrium, the constant K is also referred to as the radius distribution (R_D) of the solute in the extraction solvent and is independent of the rate at which the distribution of the solute is achieved in the mixture, but it is simultaneously affected by the experimental conditions such as physical or chemical properties related with the solvent or the solute, respectively.

Experimentally and to achieve optimum separation conditions, the analyst may optimize the best experimental conditions by making sure that the distribution of the solute between solvents A and B lies far to the right in Eq. 3.2, and the resulting values of K_D will be large, indicating a high degree of solute extraction from phase A into phase B.

Conversely, if K_D is small, the data indicate that less amount of Y is transferred from phase A into phase B in the chosen procedure. Also, it is important to consider in the analysis what will happen if K_D is equal to 1, because this value indicates that an equivalence of concentrations exista in each phase.

Usually, water is one of the most employed solvents in liquid–liquid extraction with the objective of removing a component from an aqueous solution into a solvent such as ethyl ether, hexane, or methylene chloride (all of them having lower to no water solubility). But also, water-insoluble organic solvents such as ether, hexane, and methylene chloride may contain some undesirable traces of water-soluble components like vinegar (CH_3COOH), a carboxylic acid to large dicarboxylic acids ($HOOC–(CH)_n–COOH$).

In that case, it is possible to remove these undesirable components from the organic solvent by using water as the second solvent, which is often called a water wash. To proceed with a reliable water wash, instead of performing a large and unique extraction, the wash procedure will offer the opportunity to successfully perform several extractions by using a cheap solvent like water as the best washer choice. So then, the best results are obtained when the washing process is performed with the highly recommended use of a simple separatory funnel (see Fig. 3.3). To assure the best extract recovery through the separation funnel method, the rule of thumb for liquid–liquid extraction states that several small extractions are more efficient than one big extraction, although it is a time-consuming procedure [62].

For example, test this out as follows:

During a liquid–liquid extraction, the experimental distribution of a solute in two different solvents gives a $K_D = 5.5$.

Therefore, in the equilibrium $c_2/c_1 = 5.5$ or $(g/mL^{-2})/(g/mL^{-1}) = 5.5$.

By assuming that 2.0 g of compound *Y* was dissolved in 200 mL of water and the analyst wants to extract this amount with a total volume of 100 mL of solvent 2 (organic), let us now determine if it is enough to use only one extraction with 100 mL or it will be better to use two 50 mL extractions. Thus, the chemical distribution of compound Y in the first case is $(x/100 \text{ mL})/(2.0 - x/200 \text{ mL}) = 5.5$, where x expresses grams of Y dissolved in solvent 2 and $(2 - x)$ are grams dissolved in water. Solving for x gives the analyst $x = 1.46$ ($\approxeq 1.5$) g of *Y* in solvent A, leaving 0.5 g in the water phase (or 75% are in the organic phase, and 25% in the aqueous soluble phase). In the second case, the mix has $(x/50 \text{ mL})/(2.0 - x/200 \text{ mL}) = 5.5$; solving for x gives $x = 1.2$ g, leaving 0.8 g in the water for the first extraction. The second extraction would be $(x/50 \text{ mL})/(0.8 - x/200 \text{ mL}) = 5.5$; solving for x gives $x = 0.46$; which means that in the two 50 mL extractions, we obtained $1.2 + 0.46 = 1.66$ g in the organic phase, leaving 0.34 g in the water phase. In this example, the application of two extractions was clearly more efficient than one extraction alone.

However, four 25 mL extractions would have been even better, but there are practical limitations in handling small volumes.

3.3 Effects of the Elevation of Boiling Points in Solid–Liquid/Liquid–Liquid Extraction

When an organic substance is dissolved in either non-volatile or semi-volatile organic solvent, it causes a rise in the boiling points of the solute, which is proportional to the concentration of the substance with the extent of the rise in temperature corresponding to a colligative property related with the solvent. Thus, an equation that applies for dilute solutions and non-associating substances is used to calculate the elevation of boiling point temperature in a pure dissolved solute:

$$K_D = \frac{D_{T+}M}{C} \tag{3.4}$$

In the equation, K_D is defined as the *ebullioscopic constant* (molecular elevation of the boiling point) of a particular solvent, which is a fixed physical property; M is the molecular weight of the solute; D_T is the elevation (a change) in the boiling point of the mix compared with the pure solvent boiling point; and C is the concentration of solute in grams for 1000 g of solvent.

The correlation between variables (e.g., T, M, concentration c) is a useful method for the assessment of the efficiency of a distillation equipment during the purification of liquids because the technique depends on the differences that characterize the boiling points of a pure material and its impurities. For example, if a homogenous mixture is formed by two substances that present absolute miscibility with a distinguishable vapor pressure in the ratio 2:1 (solute/solvent), it may be necessary to have the system within an efficiency with at least seven separation plates, which gives a mixture enrichment of 27 = 128, if the concentration c of the higher boiling component in the distillate is reduced to be found in the distillation flask to less than 1% of its initial concentration. If the vapor pressure is estimated to be in the ratio of 5:1, a smaller number of plates (three) would be enough to achieve a better extraction/separation of the undesired impurities from the liquid. Nonetheless, in utilizing fractional distillation, the rejection of initial and final fractions is usual, which are likely to be rich in the content of either low and high boiling point of impurities, respectively.

Therefore, the center fraction can be further extracted (purified) through a sequence of fractional distillation [63].

3.4 Liquid Extraction

To date, solvent extraction is the most used technique for a wide scope of substances that are suspected form the chemical bulk of a raw material and that could be dissociate into single molecules, ions, metabolites, or chemical functional groups after its solubility in a solvent extraction process. Nonetheless, the solubility of these raw components is greatly dependent on either the physical or the chemical properties of the solvent in a process that yields at least two physical phases composed for:

Phase 1 The raffinate phase (rich in pure solvent) and

Phase 2 The extract phase (rich in solute).

Herein, when a relative volatility is achieved in phase 1, the separation of insoluble components in the mixture is not possible by distillation processes and when relative volatility is greater than the raffinate phase, an extraction method is used for the best separation of the soluble components in the mixture [64]. In addition, if the distillation method is too expensive to carry out a simple chemical separation, then extraction will be the best choice and to do that, a basic extraction process is drawn below as a block diagram (Fig. 3.2).

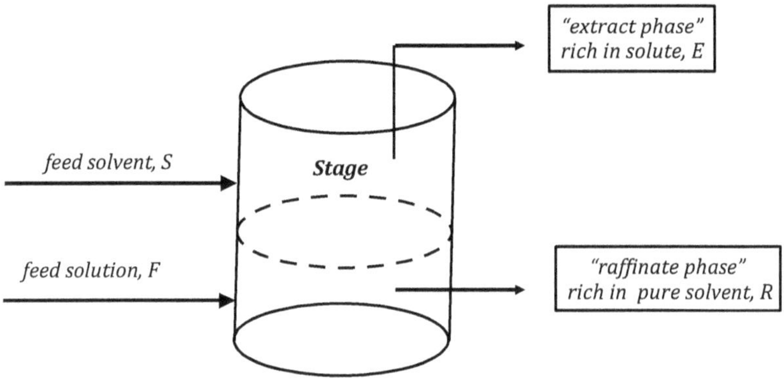

Figure 3.2 Block diagram of a single-stage solvent extraction.

Nowadays, the major challenges encountered in the extraction of fragrances, phytochemicals, folk medicines, nutraceuticals, natural dyes, and foods through the use of conventional technologies have a direct relationship with long extraction time, the requirement of using highly pure and costly solvents, which can contaminate agricultural soil similar to other environmental pollutants. Because the detriment results with some traditional extraction methods, the evaporation of large amounts of halogenated solvents is associated with its limited performance by resulting in low extraction selectivity, the undesired degradation of thermolabile compounds as well as its low possibilities for scale up.

To overcome the shortcomings encountered with conventional extraction methods, new and promising extraction techniques such as ultrasound-assisted extraction (USAE), supercritical fluid extraction (SFE), microwave-assisted extraction (MWAE), pulsed electric field–assisted extraction (PEFAE), pressurized liquid extraction (PLE), and phytonics-assisted extraction (PhyAE) have been recently introduced. A detailed discussion on these promising techniques will be presented in Chapters 4 and 5.

3.5 Liquid Extraction Independent of Chemical Reaction

When a physical extraction technique is used to separate two immiscible substances, it is supposed that until a certain point, a quantitative analysis shows that at least a parallel chemical reaction is superposed on it, which depends on the chemical reaction rate (e.g., solvent/solute ratio formation) if compared with the mass transfer rate of substances, which occurs during transitional phases (liquid ↔ gas) among whether the process can be analyzed by the chemical reaction conditions. Thus, in some cases, the chemical reaction has a direct effect on the percentage of the extraction yield, which speeds up the process. As a direct consequence, the dimensioning calculations for scaling up an extraction equipment may depend on the chemical products that are possible to be formed and quantified. However, when nonchemical reaction is anticipated to occur during an extraction procedure, it is sufficient to proceed with the separation of one or more solutes in the mix through a conventional solvent extraction equipment with the most affordable equipment to perform it laying in the use of a relatively non expensive labware instrument such as a glassware separation funnel (see Fig. 3.3), which had been designed to add and withdraw the respective liquid phase of interest.

Nonetheless, the efficiency of an extraction method needs to be quantified and, therefore, the method can be classified on the bases of two different measurable categories [65, 67].

3.6 Separation Dependent on Thermophysical Coefficient

3.6.1 Liquid–Liquid Extraction: Sample Phase (Liquid)/ Extract Phase (Liquid), Basis for Separation (Partitioning)

This extraction/separation method is carried out at room temperature and 1 atm pressure. The liquid–liquid extraction is a process that occurs as a mass transfer between a liquid considered as the solute, which dissolves into another liquid, and a solvent, which differs chemically from the former one but as a mixture forms a partially miscible or two immiscible phases defined by a thin film of contact between them. These two phases are the result of their quite different chemical compositions, which leads to a separation of individual components according to particular physical distributions or due to the molar partition between the phases. In the mixture, one phase is classified as the nonpolar organic phase while the other is the polar phase (or water). Nonetheless, this extraction method is totally different from distillation, because in the later method, the liquid solute is partially vaporized together with the solvent to create another (vapor) phase, but as steam, the two phases are similar chemically.

3.6.2 The Batch Process: Technology Dependance

The batch method is the simplest extraction procedure possible as well as a technique most employed in the laboratory for analytical separations. It involves the bringing of a given volume of solution to have contact with a defined volume of solvent until a chemical equilibrium is achieved.

The equilibrium is inferred to occur when a separation funnel is used to attain the extracted compounds into the solution, which is separated into two layers. Where necessary, the procedure must be repeated by using the same or different solvent after the addition of fresh solvent does not offer visually substantial changes in the layer's appearance (color, oil presence). Hence, the batch extraction technology operates as a rapid, simple, and clean separation process, especially if the distribution ratio of the solute of interest is in large amount whether compared with the extraction solvent. In these cases, a few extractions will be necessary to affect quantitative separation. Although various extraction technologies for increasing the distribution ratio and molecular selectivity are commonly in the market, the most widely employed apparatus to perform a batch extraction is a separatory funnel. This equipment is a relatively simple lab instrument designed to add and withdraw the respective liquid phase of interest. The extraction from a heavier liquid in a lighter solvent, it is necessary to remove the lower phase from the funnel after each extraction before removing the extracting solvent as in the case of ethyl ether extractions from an aqueous solution. Also, during a batch extraction procedure, it is necessary to follow a few simple steps that allow separation of different phases for sampling as subsequent processing for product estimation through mathematical analysis. Most batch extractors are separatory funnels designed with an upper taper off and a narrow bottom with a sealed stopcock, as depicted in Fig. 3.3.

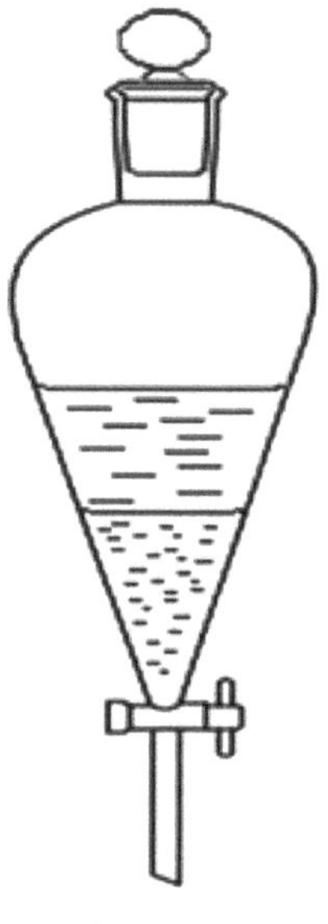

Figure 3.3 Glassware separation funnel with its simple parts and two distinct liquid phases.

The handling simplicity of this apparatus allows anyone to separate a mixture of solutes and solvents into at least two different liquid phases by withdrawing them for further analytical processing. It is, of course, essential to wait until the liquid phases have completely settled out after manual or automatic agitation. The settling step usually occurs in a matter of a few minutes. If only aliquots of the phases are to be used, it is necessary to notice any volume changes in the phase's appearance, because the mutual solubility of solute together with some impurities is proper for the solvent [68].

3.6.2.1 The extraction funnel: an empirical method

When two liquids are mixed in a separation funnel, the process is called liquid–liquid extraction (LLE), in which the extraction can be accomplished by shaking the mixture (an aqueous and an organic liquid), which will eventually separate in two phases in the separatory funnel.

Once a uniform mix is formed in each agitation step, the two layers at stand can separate by settling. The settling process starts as a flow from the bottom of the separatory funnel, which is controlled by a fritted glass or a Teflon stopcock, and the top of the funnel is sealed with a stopper (glass, Teflon, or rubber). The stopper and the stopcock must fit tightly in regards the extraction achieve not only a qualitative but also a substantial quantitative production to be a leakproof of the process. Hint! *If the desired product is a liquid organic compound (hydrocarbon, ester, light oil, flavonoid, benzenoid, flavone, heterocycle, amine), the recommended stopper and stopcock units may be constructed as fritted glassware that must be lubricated with a special grease. Conversely, a separation funnel may be assembled with Teflon material for both stopper and stopcock whether food, biological, or pharmaceutical compounds are the product of extraction interest.*

In terms of shapes and dimensions, most separatory funnels are globe, pear, cylindrical, or conically shaped to accommodate not more than one liter of capacity. They can be agitated mechanically for industrial application but are often shaken manually during research or educational sessions. With the stopcock closed and the funnel suspended on an iron ring, the liquids are poured inside the vessel, which initially form two phases when added to the separatory funnel.

The stopper is placed on the top of the funnel mouth, and the funnel is removed from the iron ring and carefully inverted without agitation. Once inverted, the stopcock is opened to relieve an excess of pressure whether the organic liquid is suspected to have low boiling point and high vapor pressure. Before the funnel is inverted, the stopcock side must be pointed away from yourself and others close to you.

The funnel may be held securely with the stopper resting in the palm of one hand, while the index and the thumb fingers of the other hand are holding the stopcock valve to open it and prevent from being blown from the funnel by pressure buildup during the iron ring removal or shaking steps.

With the separatory funnel inverted, the fingers that hold the stopcock will close and open it in a sequence of gently shaking/settling at intervals that allow for pressure release by opening the stopcock until no more steam has being generated.

If pressure builds up less rapidly in the separatory funnel, the mixture must be shaken vigorously for a longer period possible to ensure adequate miscibility of solutes while venting the steam through the stopcock occasionally. By settling the funnel on the iron ring and after removing the stopper, the layers are completely separated; the lower layer should be drawn out through the stopcock, and the upper layer should be removed through the top of the separatory funnel. The relative position of each layer depends on the relative densities of the two immiscible phases. In a batch extraction process, all layers formed may be saved until the desired product is isolated. Thus, physical chemistry properties of a given liquid layer can be easily determined as hydrophilic (aqueous soluble) or hydrophobic (organic soluble) by testing its solubility if a few drops are added to water [69].

3.6.2.2 Extraction in a continuous process

When a solute A in a liquid solvent B presents a mix distribution ratio relatively small, continuous extraction is the method particularly employed for many batch extractions that would normally be necessary to achieve a quantitative separation of A from the material in which it is contained. In that case, most continuous extraction devices operate on the same general principle, which consists of distilling the extracting solvent from a boiler flask and condensing it on a cool surface (a condenser) with a hot solvent passing continuously through the material where the solute of interest is necessary to be extracted. In these processes, the extracting liquid separates into two phases with the liquid phase flowing back into the receiving flask, where it is again evaporated and recycled with or without a fixed frequency, while the extracted solute remains in the receiving flask because the solvent density increases due to the presence of solute in the liquid extract. If the solvent is lost or cannot be easily distilled, it will be necessary to supply the system with fresh solvent to maintain the process under continuous operation for achieving the best efficiency.

In this regard, the efficiency achieved with continuous extraction depends on the viscosity of the phases formed and other physical factors that could affect the rate of attaining the equilibrium, among the value of the distribution ratio and the relative volumes of the two phases together with other operational factors. Continuous extraction is, therefore, broadly practiced in advanced organic chemistry courses, phytochemistry, combinatorial chemistry research as well in pharmacy and biotechnology scaled processes. This practical method is widely used to improve extraction efficiency by ensuring as high as possible the area of contact between the solute and the solvent to form a biphasic state created along the reflux process. During the process, as the extracting solvent passes through the extractable material contained in a fritted glassware disc (a thimble) or into a cellulose membrane thimble, or supported in small orifices (silica), baffles (thin flat objects that promote crossflow), stirrers may be used to bring the two immiscible layers into close contact.

In continuous extraction, for example, dissolved chemicals perform differently in the extracted solution, which depends upon the chemicals' solubility distribution ratio, which for practical applications is directly related with the difference of polarities presented by the individual components that form the two immiscible fluids. As discussed in the liquid–liquid extraction section, the feed solvent is present in the same phase as the extract solutes with the difference that both phases possess different density, viscosity, and polarity; they are found in different concentration and volume amounts.

Nonetheless, using continuous extraction in a liquid–liquid process, both liquids (the feed and the solute) form either a homogenous or a heterogenous mixture, which can be separated from the mix using the separator funnel as shown in Fig. 3.3 by contacting them with another liquid, which separates one of the two, preferentially under the extractant physical properties that match with the same properties of the target solute, as proposed in Fig. 3.1.

For example, if a solution of ethyl acetate (H_3C–CO–O–CH_2–CH_3) and water (H_2O) is shaken with a liquid such as acetic acid (H_3C–COOH), some portions of acetate, but relatively little water, will dissolve to form an ester phase (R–CO–O–R′) where R is the hydrocarbon part of the carboxylic acid and R′ is the alcohol group. After shaking the mix and settler to achieve equilibrium, the relative densities of the aqueous and ester layers are quite different, so they will appear to settle after the shaken stop and decantation of one phase is allows for further isolation from the layers they form. The new ratio between acid and water in the formed ester layer is deeply different from that in the original solution, which is also different from that found in the residual water solution, and a certain degree of separation between the layers is not difficult to perceive. It is an example of stagewise contact/separation of liquids, which could be performed either in a batch (using a separation funnel) or as continuous extraction. However, to improve the separation of phases in this example, the residual solvent (water) is repeatedly extracted with ester to reduce the acid content that remains in the mixture; otherwise, a countercurrent cascade stage like a Kuderna-Danish column (Fig. 3.9) could be used for this extraction process. Another possibility will be the use of some sort of countercurrent continuous-contact devise as an extraction procedure, which may be divided into three steps—i.e., extraction, cleaning, and collection—where discrete stages are not involved (Figs. 3.4–3.6).

3.6.3 Leaching or Solid–Liquid Extraction: Sample Phase (Solid), Extract Phase (Liquid), Basis for Separation (Partitioning)

Solid–liquid extraction (leaching) means the removal of constituents from a mixture of solids by bringing the solid material into contact with a liquid solvent that dissolves either a particular or various chemical constituent(s) [70]. Leaching may be used either to produce a concentrated solution of a valuable solid material, or to free an insoluble solid from a soluble material with which it is contaminated. An everyday example of leaching is *making coffee*. In this homemade process, the soluble constituents, i.e., the coffee proteins, polyphenol antioxidants, natural dyes, and organic minerals are separated from the insoluble coffee grounds by taking advantage of the water solubility of these ions and molecules [71].

3.6.3.1 Mechanism of leaching

The leaching method involves two main steps, which can be explained as follows:

A. ***Contacting step***: Herein, a solid or gel material is soaked with a selective solvent, which dissolves the soluble solute(s) in the most solvent possible. To that, the solute is first dissolved from the solid or a gel surface, and then it passes through the main body of the solution by a diffusion mechanism together with the solvent. After this initial process over a large area of the surface, it may result in the formation of cavities on the solid or gel material, which exposes new internal surfaces to subsequent solvent penetration to that surface containing solute(s) [72].

B. ***Separation step***: In this step, a separation of insoluble phases occurs. That is, the separation of the liquid phase from a solid into two physically defined phases is possible by settling, filtration, or other common methods.

3.6.3.2 Leaching operation

I. **Single Stage for Leaching Extraction**

Consider a theoretical extraction stage setup with the following parameters:

$\mathbf{F}$ = mass of A + C as solids to be leached.

$\mathbf{Y_F}$ = mass of C per mass of A + C in solids to be leached.

$\mathbf{R_0}$ = mass solution A + C recovered in the leaching solvent.

$\mathbf{X_0}$ = molar mass of C / mass of A + C in the leaching solvent.

$\mathbf{E_1}$ = residual mass A + C in leached solid.

$\mathbf{Y_1}$ = mass C/mass of A + C in leached solids.

$\mathbf{R_1}$ = mass solution A + C in overflow

$\mathbf{X_1}$ = mass C / mass A + C in the leach solution (overflow) and the overall material balance: $\mathbf{F + R_o = R_1 + E_1}$

Therefore, components in the total material mass balance equate to be:

$$\mathbf{F \cdot Y_F + R_o \cdot X_O = R_1 \cdot X_1 + E_1 \cdot Y_1}$$

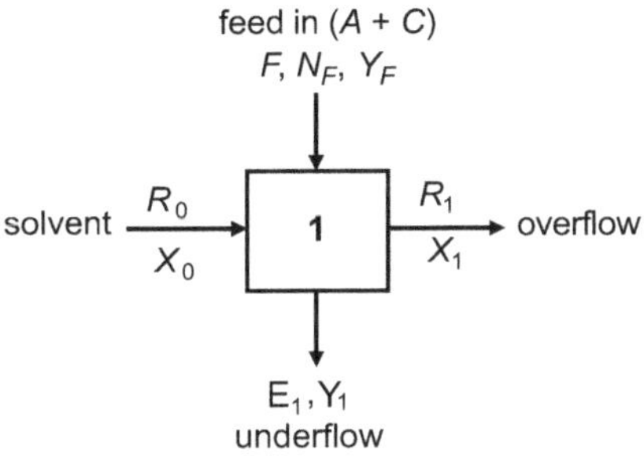

Figure 3.4 Schematic diagram of the first stage of leaching extraction. Adapted from Ref. [72].

Thus, the material balance presented in the leaching stage 1 method is too similar as that observed on single-stage material balance separated in one-batch extraction/separation process.

II. Multistage *Crosscurrent* Leaching Extraction

In the multistage crosscurrent process, the material balance in the first stage is similar to that presented in a single-stage extraction or the whole material balance as observed in Scheme 1 of Fig. 3.5.

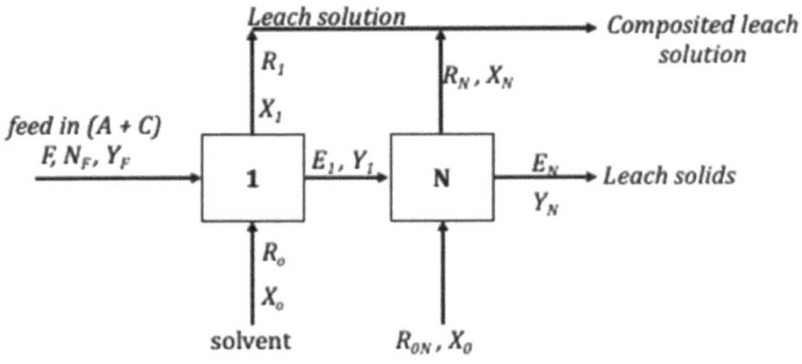

Figure 3.5 Schematic diagram of multistage *crosscurrent* leaching extraction. Adapted from Ref. [72].

Thus, the material balance of stage N on II presents an overall material balance composed by

$$E_{N-1} + R_{ON} = E_N + R_N$$

Therefore, the total component of material balance in a multistage crosscurrent extraction is found to be

$$E_{N-1} \cdot Y_{N-1} + R_{ON} \cdot X_O = E_N \cdot Y_N + R_N \cdot X_N$$

III. Multistage *Countercurrent* Leaching Extraction

In the extraction of solutes running on a multistage countercurrent system, the material balance for the N stage and the overall material balance in the process is summarized as

$$E_{N-1} + R_{N+1} = E_N + R_N$$

With the total components in the material balance expressed as

$$E_{N-1} \cdot Y_{N-1} + R_{N+1} \cdot X_{N+1} = E_N \cdot Y_N + R_N \cdot X_N$$

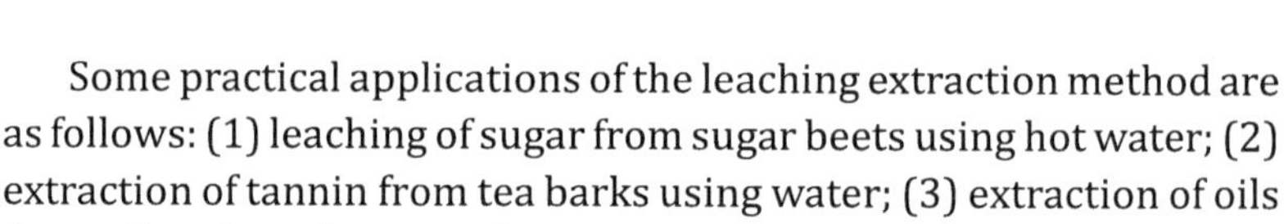

Figure 3.6 Schematic diagram of countercurrent multistage leaching extraction. Adapted from Ref. [72].

Some practical applications of the leaching extraction method are as follows: (1) leaching of sugar from sugar beets using hot water; (2) extraction of tannin from tea barks using water; (3) extraction of oils from oilseeds such as soya beans using hexane or petroleum ether as solvent; (4) extraction of fragrance oils from flowers; (5) extraction of medicinal compounds such as fine pharmaceuticals from plant (e.g., roots, leaves, stems) [73–76] and soil sample containing biosolids in the pharmaceutical industry [77]. For example, careful studies on the fate and transport of pharmaceutical residues introduced as biosolids in agricultural soil produced robust data from experiments carried out by the extraction of pharmaceutical contaminants (carbamazepine, diphenhydramine, fluoxetine, diltiazem, clindamycin), and metabolites such as carbamazepine-

10,11-epoxide and norfluoxetine as contaminants commonly found in biosolids separated from soil. The leaching experiments with these selected pharmaceuticals indicated that they presented low mobility in the tested soils. Nonetheless, small portions of these pharmaceuticals were recovered in the leachates, likely attributed to its column sorptivity if compared to their solubility in the soil's organic matter. The conducted dissipation experiments showed that carbamazepine, diphenhydramine, and fluoxetine were the most persistent in soil, while the dissipation of diltiazem and clindamycin was interpreted to be affected by chemical reaction conditions among proper chemical properties of soils that lead to redox reaction products [77].

3.7 Extraction Dependent on Equipment

3.7.1 Soxhlet Technology: A Solid–Liquid Hot Extraction

Basic techniques for continuous extraction of animal fat, oil from flower, leaves, seed, and fruit matrices are based primarily on the selection of solvent, heating medium (water, oil, sand, mantle), and agitation (mechanic or gas flow). Soxhlet extraction, particularly, has been the oldest method of solvent extraction, one of the most referenced techniques for evaluating performance and efficacy over other solid–liquid extraction methods with a few exceptions on restricted fields of applications such as the extraction of thermolabile compounds [78]. In the Soxhlet method for the extraction of lipids from algae (macro, micro), seeds, and coffee ground, for example, the process consists of separating a soluble component (oils) from the insoluble vegetable solid matrix in which it is dispersed by using an organic solvent [78]. This process is also termed "leaching" and has been previously referred to as a solid–liquid extraction to distinguish it from liquid–liquid extraction, in which a liquid organic solvent (commonly a nonpolar) is used to extract a liquid component (the solute) from another non-miscible liquid phase, usually an aqueous system [79]. The chemical mechanism that explains the solid–liquid extraction is based on the preferential dissolution of the solute, following the "likes dissolved likes rule," which in the case of oil extraction from oilseeds is already in the liquid state, from the solid matrix into the liquid solvent [80]. Nonetheless, some physical properties regardless surface area and porosity of the solid matrix, the solvent's viscosity and its diffusivity into the feed matrices, temperature of extraction and the mechanism of agitation which might be employed to improve diffusivity is described by Eq. 3.5:

$$\frac{dM}{dt} = k \cdot A(C_{\mathbf{Ssat}} - C_{\mathbf{s}}) \quad (3.5)$$

where M is the mass transferred in time t, S_{sat} is the saturated concentration of solute in the solvent, C_s is the concentration of the solute dissolved in the solvent at any time t, A is the matrix total area of contact (m^2), and k is the mass transfer coefficient [81]. Thus, the highest possible solute concentration in the solvent is the saturation concentration; therefore, the solvent-to-solid ratio must be high enough to ensure that the resulting solution will be below the saturation concentration of the solute when the equilibrium is reached [82]. The extraction equilibrium is achieved when the solute concentration is equal in both phases, the solid and the solvent phase, and consequently multiple extraction stages with fresh or recycled solvent are required to obtain high yields of concentrated extract (solute) in a time frame. At that point, high solubility of solute is achieved, and the continuous feeding of solvent will reduce the number of stages needed to reach a desired degree of solute's removal when a continuous process is applied. Regarding the chemical processes involved in the solvent extraction of edible oils, the nonpolar oil molecules are initially desorbed from their site (matrix) into the oil-bearing material in an endothermic reaction with the solvent diffusing through the organic part of the cell matrix to reach a matrix–solvent interface [83]. Thereafter, oil molecules (lipids like) diffuse into the solvent phase and more energy is required to dissociate the solvent molecules from the solute ones, so that they can accommodate them with polar solvents requiring more energy than nonpolar due to stronger intermolecular repulsion between oil and solvent molecules interacting in an exothermic process, where the energy released is higher when polar solvents are used [84].

In general terms, oil molecules dissolve fairly in a solvent where the intermolecular forces of attraction between solute (oil) and solvent molecules can be overcome by the intermolecular forces generated by the oil–solvent solution [85]. In this chemical equilibrium and based on the strength of the chemical interactions between the dissolved oil with the oil-bearing matrix, the solvent extraction process can be separated into two stages.

The *first stage* is characterized to be dominated by a brief solubility-controlled phase in which the oil–matrix interactions are quite weak, and the *second stage* is characterized to be the longer stage where desorption-controlled phase follows either strong interactions between oil–solvent phase with matrix containing oils, or long diffusion paths inside the sample matrix that leads to oil's desorption. In this regard, the Soxhlet-assisted solvent extraction has been used as the baseline method in most studies on the extraction of oils, and in most cases, when compared with other methods, the ratios obtained through other technologies are expressed relative to the yield obtained through the Soxhlet method using hexane as the standard solvent [86]. As expressed above, the Soxhlet-assisted solvent method for continuous extraction involves (1) the solubilization of oil from vegetable or biological origin in a liquid organic solvent and (b) its separation from the solvent-wetted grounds through numerous cycles of evaporation and condensation of that solvent. N-hexane [H_3C–$(CH_2)_4$–CH_3] is regarded as the most effective and less hazardous solvent in comparative studies, which have taken in consideration a broad range of organic solvents [87].

A general consensus on food, nutraceuticals, pharmacy, and phytochemical extraction efficiency agrees on the extensive use of hexane as a solvent for a number of oil extraction because this solvent possesses peculiar characteristics such as excellent oils miscibility, distinguishable boiling point range (63–69°C), and easy recoverability. However, the Environmental Protection Agency (EPA), USA, has categorized n-hexane as the topmost hazardous solvent among air pollution causing solvents [88]. This is also one of the challenges experienced by Soxhlet extraction.

Figure 3.7 shows a general picture of a standard Soxhlet system in which the seed material (solid) is either placed in a sintered porous thimble holder, in a cellulose thimble, or glued in a glass wool. The extractor vessel is continuously filled with condensed fresh solvent from a distillation flask through a lateral pipe. As the liquid reaches an overflow level inside the extractor vessel, a siphon aspirates the solution of the thimble holder and unloads the solute–solvent phase into the distillation flask, carrying the extracted solutes into the bulk of the liquid. Solutes are left in the boiling flask, and fresh solvent passes into the solid bed several times with the operation repeated until complete extraction is achieved.

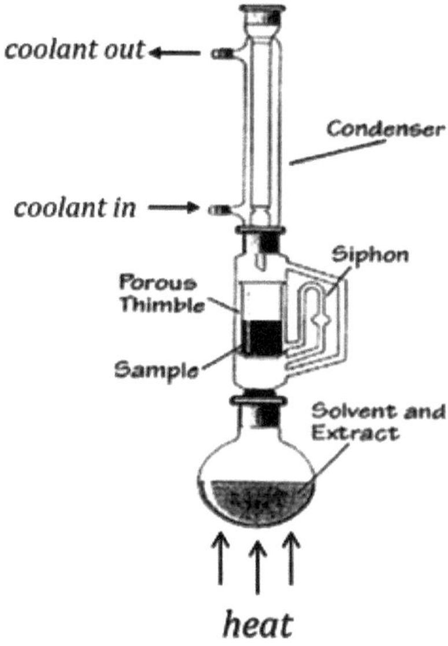

Figure 3.7 Scheme of a Soxhlet system setup for solid–liquid extraction.

Once the extraction is considered complete, the flask containing the extracted materials (solute) is removed from the extractor vessel and the solutes are separated from the solvent using evaporation through a roto-vapor system or using a separation funnel and different solvents as discussed in Section 3.6.2.1 (Fig. 3.2). However, we clarify at this point that *Soxhlet extraction* and *heat reflux extraction* are not the same process.

First at all, heat reflux extraction can be performed simply by placing the feed material in a boiling flask together with the extraction solvent. Over the boiling flask is surmounted a chilled surface (a condenser) to condense the rising solvent vapors as they boil off and return them to a liquid state into the refluxing flask, without boiling away (Fig. 3.8). Continuous extraction is particularly applicable when the distribution ratio, K_D, between the solute and the solvent is relatively small, and then many batch extractions would normally be necessary to achieve a reliable and quantitative extraction of solutes from a material.

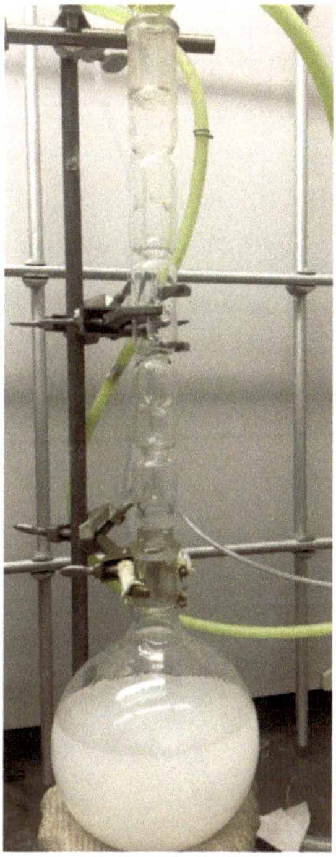

Figure 3.8 Typical heat reflux extraction system (refluxing flask bottom and Kuderna–Danish column). Courtesy of author's lab.

Most continuous extraction devices operate on the same general principles, which consist of distilling the extracting solvent from a boiler flask, creating a vapor (or steam), which condenses and hot liquid passes continuously through the solution (or molecules) being extracted.

The extracting hot liquid separates out the solute from its matrices forming the "extract" solution, which flows back into the boiling flask, where the solvent is again evaporated and recycled while the extracted solute remains in the refluxing flask. If the solvent cannot be easily distilled, a continuous supply of fresh solvent may be added from a reservoir attached to the boiling flask. Moreover, the extraction efficiency in a continuous process depends largely on the viscosity of the phases formed among other physical factors affecting the rate of solute extracted to attain equilibrium with the value of the distribution ratio and the relative volumes of the two phases and other factors analyzed through Eqs. 3.2 and 3.3.

A practical method employed to improve extraction efficiency is enhancing, as high as possible, the area of contact between the solvent and the solute. Thus, as the extracting solvent passes through the material containing solute, fritted-glass discs, small orifices, baffles, and stirrers may be used to bring the two immiscible layers be in closer contact. Several engineering devises that afford for continuous extraction or solvent purification had been designed with different features for specific requirements (see Fig. 3.8).

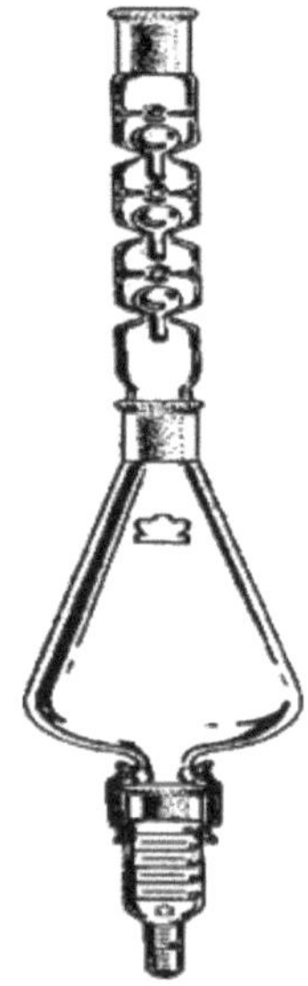

Figure 3.9 Scheme of a reflux system for solvent mix purification termed Kuderna–Danish extractor concentrator.

All these extractors have been designed to operate in continuous solvent extraction regimes; by doing that, the components of interest (the solute) are dissolved in the liquid mixture, which can be separated from the fluidic contact between the solvent and the solute with an insoluble liquid that preferentially dissolves the components that are part of the sample's organic structure according with some selectivity or specificity rules (e.g., likes dissolve like or polar solvents dissolve either polar and ionic solutes), which apply for extraction but are subjected to the solubility rules discussed above.

3.7.1.1 Benefits and drawbacks of Soxhlet extraction

A general consensus agrees on the experimental and profit benefits achieved by using conventional Soxhlet extraction, which includes (1) keeping the system far from equilibrium by constant exposure of the solid matrices to fresh solvent percolation, (2) maintaining high extraction temperature to enable recovery of the compounds of interest, (3) not requiring filtration of solute(s) after leaching, and (4) requiring a continuous process with less to no attention for long periods. In addition, the process is very simple and classified as a low-cost technique [88]. The major disadvantages encountered using conventional Soxhlet extraction method include the following: (1) The extraction time is lengthy and the final step is labor intensive due to disassembling of the system, solute(s) recovery from the vessel, and solvent evaporation. (2) A considerable amount of solvent is consumed. (3) Because of the extractor design, agitation cannot be provided in standard devices that allow to speed up the process. (4) Due to a large amount of solvent used, it needs an evaporation/concentration procedure that takes long times. (5) The high temperature applied to speed up the process presents a high risk of thermal decomposition of thermolabile target compounds. (6) The process cannot be performed in a selective extraction fashion. (7) The process does not allow to manipulate variables as data useful for mathematical analysis. (8) The combination of long time and the requirement of a large amount of solvent results in two major limits for the scaling up of the Soxhlet extraction technique [89].

3.7.2 Soxtec Technology: A Soxhlet Accelerated Solvent Method

The Soxhlet procedure remains the most exhaustive extraction technique, and today it is still widely used. Nonetheless, the evolution of the Soxhlet extractor had been critically discussed, and the conclusion from a large number of revised works is that the adoption of new technologies to improve its performance has converted the Soxhlet method to almost a "panacea" in the extraction field. Nowadays, by assisting the method with high pressure, ultrasound, or microwave technologies, it has decreased or minimized the negative concerns associated with the conventional extractor design.

Thus, the automation of Soxhlet performance has opened the door to commercialization of a few different innovations [89]. Special attention to innovate on the method relays with its efficiency, and there had been some improvements in the basic technique, albeit for profitability, where the conventional procedure remained long, tedious, and prone to variability.

A comparison of the extraction efficiency of diverse materials by using Soxhlet and a Soxtec technologies is shown in Table 3.1. The time consumed using the conventional method (Soxhlet) versus the modified extractor (Soxtec) depicts deep differences that, in a practical sense, are related with investment versus revenue. Nearly a century after the Soxhlet invention, in the early 1970s, Edward Randall [90–92] developed an accelerated extraction (ASE) technique aimed to reduce the extraction time to as little as 30 min when compared with the traditional Soxhlet extractor.

Table 3.1 Soxhlet versus Soxtec extraction time comparison for diverse materials using same solvent(s)

Analyte	Soxhlet Extraction Time	Soxtec™* Extraction Time
Fat in feed	2 h to 16 h	1 h
Fat in food (e.g., meat, cereals)	2 h to 16 h	1 h
PCB, PAH, and pesticides in soil	17 h	2 h
Oil and grease in water	4 h	45 min
Extractable matter in polymers and rubber	2 h to 48 h	1 h to 5 h
Extractable matter in paper and pulp 2 h to 48 h	1 h to 5 h	
Hydrocarbons in petroleum rock	24 h	2 h
Finishing oil on textile and synthetic fibers	2 h	30 min
Anti-caking coating on fertilizers	3 h to 4 h	1 h
Detergent in detergent powder	5 h	1 h
Fat in leather	5 h	1 h

PCB, polychlorinated bisphenols; PAH, polyaromatic hydrocarbons.
Adapted from Ref. [92]: *FOOS.com. Soxtec™ 200 series Solvent Extraction Solution.

In the Randall-improved vessel, the sample is lowered and totally immersed in the boiling solvent, contrary to Soxhlet extraction where the sample is in a static position, exposed to an either warm or a cold solvent that comes in contact with the extract in cyclical intervals. In the new procedure, the materials that will be submitted for extraction (e.g., fats, oils, fragrance molecules, or waxes) become more soluble in hot solvent than in cold one (traditional Soxhlet) or extraction processed with the solvent acting on the extract at room temperature (e.g., maceration).

The Randall method includes two steps to process an extraction, which begins with a boiling step, followed by a rinsing step to flush the residual extract from the sample (Fig. 3.10). Because of the excellent agreement with the continuous Soxhlet performance, the improved precision presented by the Randall innovation makes it the inspiration to design and develop many other automated systems in the last three decades.

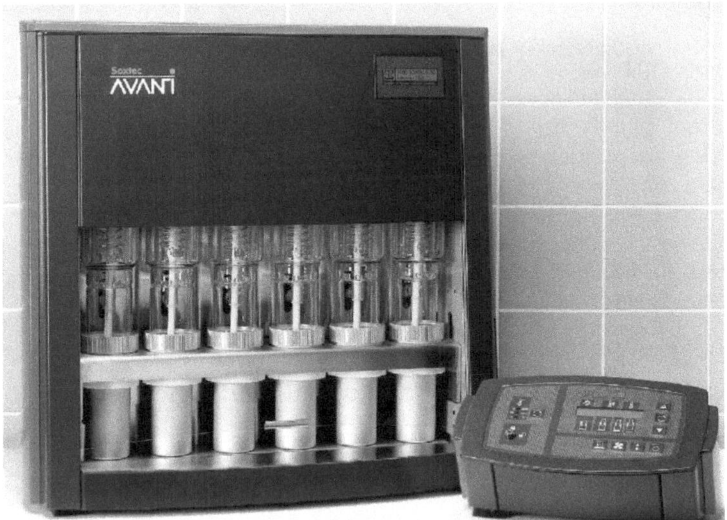

Figure 3.10 The Soxtec™ Avanti 2050 automated extraction system. Adapted from Ref. [92].

Shortly after the modified Soxhlet invention was patented, in 1975, Tecator AB of Höganäs Sweden acquired the US patent rights to what had become known as the "Randall Modification Extractor" for the Soxhlet method. Initially, it was first commercialized as the

RaFaTec and later became the Soxtec systems as are known today. Figure 3.11 shows a diagram of the former design of the Soxtec™ Avanti, from Foss Tecator, an automated Soxhlet extraction system using the Randall submersion technique.

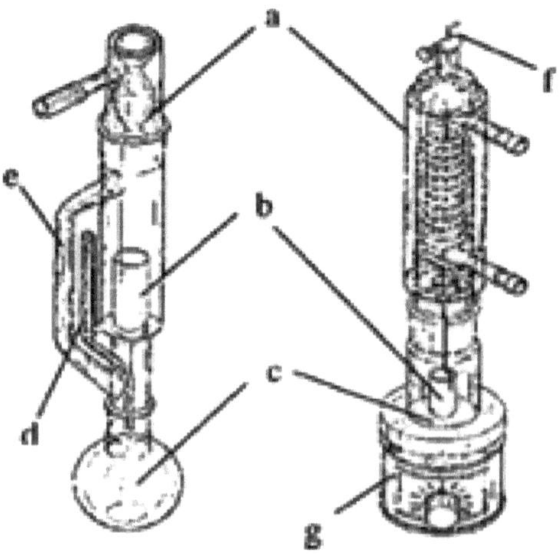

Figure 3.11 Modified solvent extractor, with the original Soxhlet (left) and Randall-modified apparatus (right) where (a) condenser, (b) sample thimble, (c) solvent refluxing flask, (d) siphon tube, (e) solvent vapor tube, (f) thimble lift/downward positioning mechanism, (g) heater (not shown on the Soxhlet). In the original Randall method, the thimble is positioned by the use of the slide rod (f). Lowering the thimble (b) into the boiling solvent for the boiling step, then raising it out of the solvent for the rinsing step. In both stages, condensed solvent is flowing continuously through the sample and thimble back into the boiling solvent. Adapted from Ref. [92].

3.7.2.1 Soxtec operation: automation of Soxhlet extraction

The Soxtec automated extraction method has gained widespread acceptance in research as well as analytical protocols and has been accepted for several regulatory agency approvals worldwide. The method is currently used on hundreds of different sample states for a broad scope of chemical extracting solvents. Herein, a brief discussion is presented on fast crude fat extraction. Crude fat extraction is assisted by solvent methods and had been classified as an empirical process [93], which means that the result will depend on both the chemical properties of the solvent(s) and the physical variables set up to run the method.

Nonetheless, it becomes critical that parameters applying to the procedure be followed strictly for data reproducibility and extraction accuracy.

3.7.2.2 Soxtec efficiency: sample preparation dependance

A proper handling of sample and attention to minimal preparation details are extremely important for the success of an analytical process (accuracy, reproducibility, repeatability). For example, an improperly or sloppily prepared sample leads to invalidate even the most carefully performed extraction procedure. As a direct dependency on either type or nature of sample, its previous preparation may incorporate different procedures adjusted to fulfil the process protocol. For example, for grinding and weighing, it is recommended that the sample shows homogeneity and be finely ground, usually to pass through a 1 mm sieve (~18 mesh). In this regard, a particular attention may be paid to the type of grinding mill used. Thus, the mill or milling process should not contribute to any loss of moisture or fat mass from the sample treated. Related with the mass, every sample must be weighed using an analytical balance calibrated to record at least four decimal place digits, with the sample weighed directly into an either cellulose or sintered-porcelain thimble. In the current example, the weighted sample is exclusively dependent on its approximate fat content. Thus, Table 3.2 is used herein as the example's guideline. Because a few grams of sample are used for analysis in the example, it is critical that this small amount be considered sufficient and representative from a larger sample's lot.

Table 3.2 Soxtec expected fat content and sample weights

Fat Content (%)	Sample Weight (g)
0–10	2–3
10–25	1–2
>25	0.5–1

Reproduced from Ref. [92].

3.7.2.3 Sample pretreatment for Soxtec extraction

Drying: For the fat extraction, it is recommended that most samples should be pre-dried to optimize the process, because the presence of water in the sample can decrease the efficiency of the solvent during the extraction, resulting in low fat yield recoveries. If the sample is rich in high water-soluble components such as carbohydrates, glycerol, salts, and urea, they could be extracted in an organic solvent with fat, yielding falsely high recoveries, because a large percentage of these chemicals is soluble in each liquid phase.

To avoid experimental errors related with analytical miscalculations, the sample may be weighed directly into the extraction thimbles, which are typically dried at 102 ± 2°C for 1–2 h [94–97]. On the other hand, when the sample is weighed before drying, the results are reported on an as-is basis by doing the accuracy less confident [96]. For example, if the results are expressed on a dry matter basis, it must be calculated from a separate moisture determination. For very moisten samples, sand may be mixed with the sample before drying. With this procedure, the sample is prevented from becoming caked during drying with an improvement in the solvent flow through the sample matrices and affording for an optimal extraction (e.g., extraction of crude fat in meat and meat products).

Acid hydrolysis: For samples that had been processed, cooked, or extruded, they often have portions of fat remaining, which are bound to lipoproteins, carbohydrates, and/or biological minerals, making them unavailable for solubilization. Then, acid hydrolysis is applied in which, the target sample is boiled under the presence of (preferentially) hydrochloric acid, which breaks the hydroxy bonds and allows the fat molecules to be dissolved gently by an organic solvent and eventually fully extracted. The acid hydrolysis method is generally used in baked products and pet foods because it facilitates the extraction of fatty acids from glycerides, glycolipids, phospholipids, and sterol esters, which might otherwise be left un-extracted due to abundant covalent and ionic bonds present in these materials.

The pretreatment utilizing acids can also facilitate the organic solvent coextraction of additional non-lipid materials. For food and pharmaceutical samples, the suitable molecular digestor of choice is hydrochloric acid (HCl), which leads to an easy breakdown of covalent and ionic bonds on lipids, proteins, and carbohydrates; by doing that the lipids commonly bound to these fractions can be extracted subsequently by an organic solvent such as hexane. When extracted, crude fat is found in lower values than expected, especially when its origin comes from an animal food product that has been heat processed or if it is suspected that it contains ingredients that have been also heat processed, the acid hydrolysis should be considered as the pretreatment method of choice previous the material fat extraction in regards of improve fat content yield after solvent coextraction. In addition, the acid hydrolysis method is often used to improve the extraction of high fat bioproducts, such as calcium salts of fatty acids, which are analyzed for crude fat as well for emulsified fats. As a point of reference for heat-processed animal feeds that should be analyzed without acid hydrolysis pretreatment, there are those products dependent on temperature that reaches during pelleting between 160 and 185°F, which does not need acid hydrolysis for crude fat determination. For expelled and expanded products, their preparation temperatures are generally 240–280°F, with high temperature accounting for extrusion products that need to be processed at 280–325°F (e.g., dry pet foods) and baking products with temperatures between 325 and 400°F. For high-temperature extrusion processes, they are usually followed by a dryer stage, where air temperatures are set up between 400 and 600°F to drive moisture off with the help of fan(s) that are designed inside the oven chassis. In the case of processed wet pet foods (cans, trays, or pouches), they are commonly subjected to retort temperatures (240–260°F) and may require a treatment with acid hydrolysis for analysis after its initial drying is accomplished during the sample preparation step. As an example, to produce pet treats, they need to be processed at low temperature under extrusion techniques that are usually close to the expeller temperature processes (240–280°F) and may or may not require the dryer step [94].

The manufacture of calcium salts of fatty acids is a process that also requires the usage of high temperature; therefore, acid hydrolysis would be necessary for the determination of its crude fatty acids. Thus, by using spray-drying temperatures for food and animal feed processes, it may be proceeding at 275–325°F, which also justifies a pretreatment with acid hydrolysis before its crude fat extraction. On these regards, it has been though that a combination of chemical and physical pretreatments for the extraction of crude fatty acids from an either processed food or animal feed product, a consensus states that for all food products processed at temperatures over 200 °F, they may need to be carried out through an acid hydrolysis method [94].

In addition to the correlation between temperature and drying for fat extraction, most dairy products and products characterized for their high sugar content, they will require a *high pH* pretreatment process by using, for example, liquid ammonia (NH_4), because this procedure leads to an easy breaking of emulsified fats as well as proteins that are solubilized before being submitted to organic solvent extraction. In these procedures, the extraction is commonly carried out with a solvent azeotrope formed by the blend of diethyl ether and petroleum ether, which is commonly known as the Roese–Gottlieb method [94].

Sample's rinse: For a sample that contains a large number of water-soluble components, it is understood that the sample may exhibit limited to poor organic solvent extraction efficiency. To solve potential organic solvent extraction limitations laid to watering, a pre-extraction treatment with deionized water (preferably), followed by a thorough drying step, can be used to obtain an acceptable extract recovery by removing several water-soluble components. To proceed with this treatment, it is specified that in the prewashing of the weighed sample, five aliquots of 20 mL of deionized water may be applied followed by the sample to be dried on a preheated oven or through dried vacuum lane, with the organic extraction being carried out by following careful specifications [38, 39].

3.7.2.4 Soxtec accelerated extraction: solvent properties dependance

After a sample belonging to the same or different origin has been properly prepared or pretreated, it is separately placed in each Soxtec vessel for fat extraction. However, before the extraction stages act on the sample, the weight of the cleaned and dried extraction cups is carefully measured, and these data must also be obtained (saved) for later use in the calculation of the final yield of products. Then, the samples in the extraction cups are carefully positioned in the extractor system, with the solvent being added through a closed-loop addition application and the extraction begins.

The sequential steps such as boiling, rinsing, and evaporation/solvent recovery proceed in an unsupervised manner with a time set up to stop the process once the extraction is considered complete and advised by an alarm signal that has been set up as an experimental parameter.

Boiling step: In this step, the thimble filled with sample is lowered and totally immersed in the boiling solvent contained in the extraction cup. Hot solvent vapors reflux against a chiller-jacketed cooling column with condensed solvent flowing back continuously through the sample by returning to the boiling solvent (Fig. 3.12) in a continuous process. This step is considered the key aspect to accelerating the extraction process when compared with the traditional Soxhlet method. In the Soxtec method, the solvent simply solubilizes the extracted solute(s) faster in the hottest solvent, by decreasing the time required for extraction in several folds. For an optimal extraction, it is recommended that the level of boiling solvent must be higher than the sample in the thimble at least one fold.

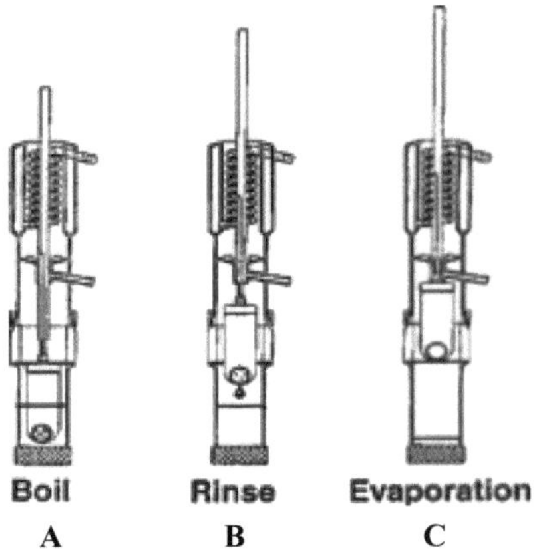

Figure 3.12 The Foss Tecator Soxtec Avanti, three-step extraction procedure automated system is based on the Randall modification of the Soxhlet technique. In the boiling (A) and rinsing (B) steps, solvent is refluxed within the condenser. During refluxing–evaporation (C), the solvent flow is blocked from returning to the extraction cup and flows out of the tube (A) into a collection tank (not shown). Source: Ref. [92].

To avoid the loss of sample material, on the top of the thimble, a medical-grade cotton piece is placed, which was carefully defatted by soaking it in diethyl ether or hexanes for 24 h, agitating several times during this period. The solvent is then removed, and the cotton is air dried. This part (cotton) is commercially available from Foss Tecator, Part No. 1529-0009. When using the Soxtec Avanti system, 70–90 mL for solvent extraction, the time frame ranges from 20 to 40 min depending on the solvent's physical properties as well as the sample characteristics. Once the boiling step is complete, the sample rinsing step begins. In this step, the sample is raised from the boiling solvent and suspended over it. In the suspended position, residuals and other chemical traces of the extractable material are flushed out by hot vapor or steam and are retained in the extracted material deposited in the cup. This step is a bit longer than the boiling one and usually takes about 10–20 min longer to ensure complete extraction [92]. The final step is achieved when the crude fat is separated through evaporation for solvent recovery. Although the condensed solvent continues to boil and evaporate in a continuous cycle, by using an internal valve, the condensate is redirected out of the condenser. The evaporation step is complete when all solvent is driven off the cup and the concentrated extract is attained. This step usually requires about 7–10 min depending on the solvent amount, density, and boiling point. A valid concern on excessive drying is associated with the possible oxidation of the product extract, which could cause weight changes among erroneous readings.

Most Soxtec Avanti model extractors store the evaporated solvent in a common collection tank for purification, cleaning, and further reuse. The Soxtec Avanti model also offers an optional fourth, cup pre-drying step, in which the extraction cups are raised a few millimeters off the heating solvent surface allowing the radiant heat to complete the final drying cycle in a heat transfer mechanism. This technological inclusion confers special applications for samples in which the targeted material(s) is(are) recognized to be extremely thermolabile.

Extracted mass calculation: The crude fat obtained by solvent extraction is calculated by a gravimetric method. Thus, the final product is calculated from the original sample weight and the weights of the extraction cup before and after the extraction are as:

$$\%\ \text{Crude fat} = \frac{W_{\text{F}} - W_{\text{C}}}{S} \times 100 \tag{3.6}$$

where W_C is the weight of the extraction cup, W_F is the final weight (= extraction cup + extract weight), and S is the crude sample weight. With all weights recommended to be recorded on a balance with ±0.1 mg (= 0.0001 g) of accuracy.

The drying effect: Considering the empirical nature around the Soxtec method, it is highly recommended to maintain a consistency on all aspects related with the extraction procedures. Thus, an often-overlooked aspect for fat extraction is to do not miss the pre-drying step of sample. As described above, any trace of water in the crude sample may contribute to increase errors in at least two ways: (1) water can act as a physical barrier that prevents solubility of fat into the organic (less polar) solvent, thus generating low fat recoveries; (2) it can also contribute to falsely give high apparent fat recoveries (false mass) represented by water-soluble components such as urea, light proteins, or carbohydrates that are co-extracted together with the fat. Because extraction had gained track as an industrial method for food analysis, in the interest of time spent and productivity costs, many laboratories do not include in its protocols the pre-drying sample steps. To correct the inconsistences of data collected on diverse samples from the same lot, the intrinsic error generated for each type of sample must be analyzed fully by carrying out extractions on samples with and without pre-drying.

By doing this, the error is cancelled whether the procedure is applied systematically on the same type of sample by setting up the extractor system with same parameters and using the same amount of sample always.

As an example, Fig. 3.13 illustrates the percentage (%) of fat recovery from dried versus undried food from different sources. The moisture detected in food samples is in the range of 5 to 25% from the total mass. In common experiments, it had been found that some samples showed clearly "water effect" more than others. In these cases, samples such as texturized feeds, which are characterized for contain molasses, and those used as feedlot concentrate that contains heavy and highly hydrolysable molecules (like urea), are examples in which they failing to a pre-dry treatment because their marked hydrolysis character which have significant effects on fatty acids extract recovery. In addition, the process is highly dependent on the purity of solvent used. Therefore, % of fatty acid extract recovery is calculated as:

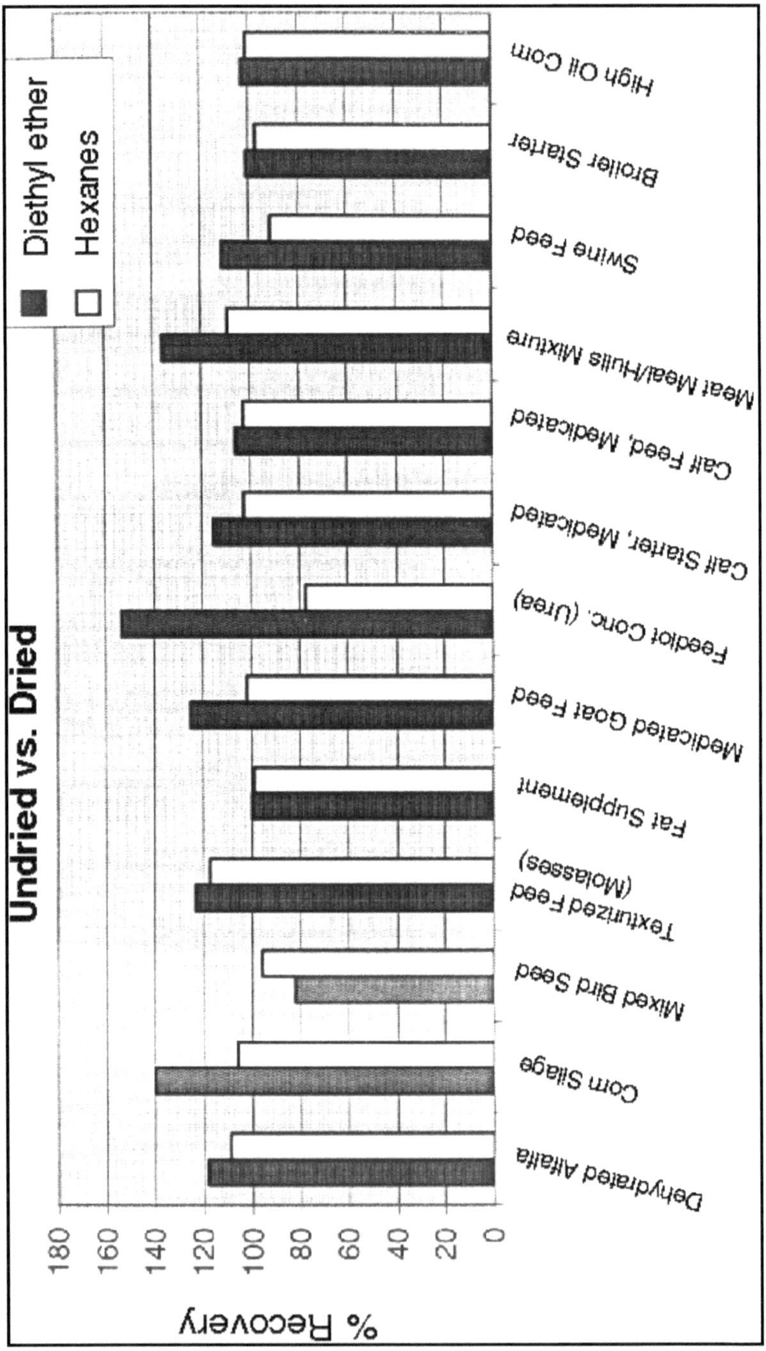

Figure 3.13 Apparent fat recovery from dried versus undried samples with moisture content ranging from 5 to 25%. Source: Ref. [92].

$$\% \text{ Recovery} = \frac{\% \text{ crude fat (hexane, ethylether) extract}}{\% \text{ crude fat from dried sample}} \times 100 \tag{3.7}$$

Solvent effects: Comparing the Soxhlet versus Soxtec versatility as extraction methods, both allow the use of various classes of non-volatile to semi-volatile organic solvents. The large list includes ethers; aliphatic, aromatic, olefinic, heterocyclics, and chlorinated hydrocarbons; and different alcohols. Because solvents are characterized to possess different solubility properties, the solvent action on a sample will have somewhat different effect for fat yields depending on the solvent characteristics. For example, for crude fat extraction, diethyl ether and petroleum ether are the most used solvents. Nonetheless, the peroxide-forming nature of diethyl ether causes it to be a less desirable choice for routine use in an academic laboratory. Thus, the explosive nature of peroxides has led to restrict the use of diethyl ether and many laboratories had substituted it with alternative solvents like fluorinated hydrocarbons.

Currently, petroleum ether had been abolished for fat extraction and totally substituted. But we need to clarify that petroleum ether derivatives or so-called ligroin are not true ethers but mixtures of aliphatic hydrocarbons and can be purchased in various formulations and boiling point ranges.

3.7.2.5 Solvent removal

The final product concentration depends on the removal or evaporation of solvents and is an essential step seriously evaluated in extraction processes for pharmaceutical, chemical, and biotechnology industries, in applications where it is necessary to reduce the solvent volume to a certain extent to facilitate formulation or analysis of molecules of interest.

Together with evaporation techniques, the use of drying or concentration methods for natural products, medicinal or chemical research, and high-quality production of flavors and fragrances is quite widespread, yet preparative purification remains their most common application.

Because the large diversity of sample types and solvents, various commercial systems have been developed over the years to accommodate the wide range of applications without a universal solution being available till now.

3.8 Solvent Evaporation: Thermodynamic Principles

The thermodynamic stability of any substance such as a gas, liquid, or solid will depend on its chemical response to changes from one state to another due to the physical effects introduced by changes in pressure or temperature operating on the container the substance is processed.

Thus, these thermodynamic effects promoted by both pressure and temperature on a defined substance are shown in the so-called "phase diagram" (see Fig. 2.6 in Chapter 2). Herein, when a solid is being heated under constant pressure above the triple point, it will reach the melting point and liquefy. Further heating will lead to an increase in temperature until the boiling point is reached, and the liquid will then start boiling, changing into the gas phase [96]. Nonetheless, when a similar work is performed with both temperature and pressure below the triple point, the material will not melt but lead the vapor to condense to a solid, instead of a liquid, in a process defined as sublimation [97]. In this case, the heat energy supplied to the sample at low pressure transfers enough energy for thawing; however, the pressure supplied is too low for the liquid formation and then, the solvent will, therefore, sublimate into the gas phase.

Because the physical phase of a substance is determined by both the heat and the pressure at the observation time, the temperature at which the material boils or evaporates is set by the exerted pressure on it. Thus, by reducing the pressure on the surface of a liquid substance or by applying a vacuum, it can, therefore, lead to a substantial decrease in the solvent's boiling point and consequently to an easy assisted vaporization, which would occur at lower temperatures. Therefore, low-pressure systems are commonly used in separation methods for heat-sensitive samples to decrease the boiling point of the substance in a mix so that the solvent vaporization occurs at a lower and, for a thermosensitive substance, safer temperature. Hence, a similar thinking can be done for sublimation processes [97].

3.8.1 Rotary Evaporation: The Rotavapor System

Rotary evaporators have been designed for quick, gentle evaporation and condensation for most common solvents, and a general scheme of a commercial rotavapor is shown in Fig. 3.14. A long time ago, rotavapor equipment had been used as rutinary tools for the safe separation of highly corrosive to very volatile solvents for mixtures of liquids or liquid-solids possessing high boiling points. So then, academic labs as well as food, biotechnology, and the chemical industry had adopted the rotavapor system for solvent removal from the final stage of a chemical reaction for the separation of mixed solvents, for the precipitation of suspensions from solutions, for concentration of liquids, in re-crystallization of a sample to remove impurities, to dry a powder or granulate samples, for synthesis of chemicals, after Soxhlet extraction, or in the separation of solvents for recycling.

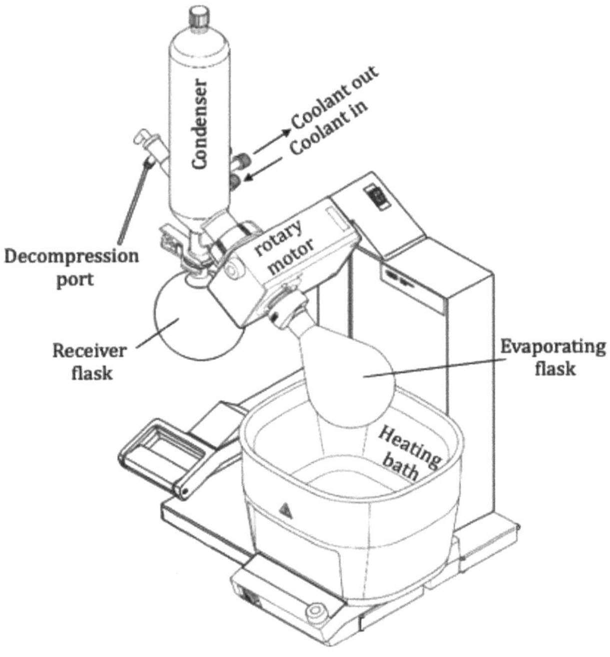

Figure 3.14 Scheme of a commercial rotavapor.

During the rotation mechanism, the flask containing the sample increases the surface area of the mixture, thereby improving the heat transfer through the sample's liquid phase and making the solvent evaporation process faster.

The constant rotation also makes the vaporization easier by avoiding localized overheating and incrustation, which occur when the sample remains static and heat is absorbed on a determinate surface area all the time. In this process, the motion of a liquid sample in both twisting and spiraling patterns, reduces its boiling time and avoids foaming by reducing formation of sticky mass of small bubbles (frothing) by doing the solvent evaporation one smooth and safe. Because the process occurs under moderate to high vacuum, it lowers the standard boiling point of solvents by making the evaporation to occur at a low temperature as an additional benefit to take into consideration in a chemical extraction analysis.

Chapter 4

Synergism and Its Complementary Effects in Chemical Extraction

4.1 Introduction

A considerable number of books and journal reviews had been written in the recent years on the importance of so-called "synergism," in an effort to provide valuable insights into the accurate definition of the combination effects derived from complex mixtures when compared with that determined by using a pure substance. Although the effects due to chemical interactions between different bioactive constituents in either medicine or nutraceutical compound have gained popularity in many research disciplines, it remains difficult to give an accurate definition for the term *synergy*, implicitly understood as a collection of effects responsible for irrefutable positive results. Thought as specific considerations, the scientific community generally agrees that the interactions between multiple agents in either experiment or process can be classified as antagonistic, additive/non-interactive, or synergistic. In this regard, additive and non-interactive combinations indicate that the combined effect generated by two substances possessing somewhat different chemical composition is pure summation effects, while an antagonistic interaction is the result of less than an additive effect. Therefore, a positive interaction between two substances can be

Semi-Critical Assisted Extraction: Applications and Commercialization in Biotechnology, Food, and Pharmacy
Tulio Chavez-Gil

ISBN 978-981-4968-36-2 (Hardcover), 978-1-003-29124-4 (eBook)
www.jennystanford.com

defined as a potentiation or synergistic effect because the presence of the two substances when combined produces effects greater than those expected from an additive effect alone.

4.2 Synergism: A Chemistry Perspective

The phenomenon of synergism was first observed in biological systems as an improved response to the chemical activity of two or more bioactive compounds when used as medicine for the treatment of a disease when compared using a single pure substance [99]. Applied to the extraction phenomena, synergism is a measured property related with an either physical or chemical method as well as combination of both, that can work on different proportions but on the same objective (the sample). Thus, when applied as single method but through a synergistic strategy (e.g., **physics**—sonication plus pressing or **chemistry**—*polar* plus *non-polar* solvents) or as combined methods (e.g., pressing—physical, plus solvent—chemical) to improve extraction or separation of chemical substances due to the phenomena appear to produce positive combined effects greater than the sum of individual and separate effects by doing its application a reliable activity which affects a methods' "**efficiency**." In last two decades, this approach has been considered an alternative for solvent extraction to curb the use of chlorinated substances (the most used in the past for phytochemistry extraction) for mixtures of acetonic and alcoholic solvents with light hydrocarbons to obtain pure azeotropes, when applied to vegetable and food extraction shown products possessing high purity and compared to their former halogenated parents, the chemical synergism of selected solvents give better efficiencies than those offered for each individual solvent used in the same process [99].

Thus, the synergistical effect in an extraction can be quantitatively described by means of the synergistic coefficient C_S, which is experimentally calculated as a function of the phase distribution coefficients of extractants or by the difference in extraction yields [99, 100]. Hence, in a chemical extraction performed utilizing two extractants designated as A and B, the synergistic coefficient

C_S represents the ratio between the distribution coefficients corresponding to the mixture of extractants, and the sum of the distribution coefficients is obtained separately for each extractant as:

$$C_S = \frac{D_{AB}}{D_A + D_B} \tag{4.1}$$

From Eq. 4.1, the coefficient D_{AB} is part of the discussion on Section 2.2 (Chapter 2) with respect to the volatility of a solute on a specific solvent and how herein the synergism effect between solvents A and B exists only if the value of CS is greater than 1. Therefore, if the calculated value is less than 1, the effect is called "negative synergism" [99]. Nonetheless, according to the literal meaning of the word, all extraction mechanisms that lead to the formation of a ternary complex can be classified as a synergistic process; however, the term "synergism" that will be employed in this chapter is restricted to a phenomenon in which the extractability of a natural compound is increased using semi-volatile to nonvolatile solvents.

For practical purposes, the main effect of synergism when applied to extraction consists of improvement in the hydrophobicity of an extracted substance produced by the reaction between the solute dissolved in aqueous solution and the extractants from the organic phase.

It implies that the synergistic effect is the result of two simultaneous phenomena:

1. A modification in the individual extraction capacity of each extractant, and
2. The capacity of the solvent mix to modify the chemical structure of the solute when compared to that exhibited by the effect of an individual extractant.

4.2.1 Solvent Effects to Consider in a Synergistic Mixture

It is assumed that one extractant (liquid or gas) reacts with the solute by forming a solute–extractant complex in the organic phase, while other part of that substance increases the complex hydrophobicity by solvation. This concerted activity is understood as the synergic effect, which in terms of chemical solvation and chemical equilibrium is amplified due to the increased difference created between the polarities of the extractants used in an extraction or separation among their respective concentration [101]. Another important parameter to consider in applying synergistic solvent approaches is their nonvolatile character and thermal stability, which can make them our best alternatives if semi-volatile to volatile organic solvents are the choice for an extraction procedure. Moreover, the blending of two different liquids during an extraction is expected to significatively stimulate chemical changes in cell walls and to influence its lipid extraction efficiency. Nowadays, the attention on chemical extraction synergism is focused specially on increasing the yield of extraction (the method efficiency), with substantial reduction in extraction steps, decrease in energy requirements, and finally, reduction in process costs [102].

4.3 Case Studies of Chemical Synergism

4.3.1 Case 1. Lactic Acid Extraction by Synergistic Effect of Tertiary Amines

An extractant mixture formed by tripropylamine (TPA) and tri-octylamine (TOA) dissolved in a weakly polar/nonpolar mix of 1-octanol/n-heptane was used in the active extraction of (L^+) lactic acid in an aqueous solution. For the same purpose, synergistic extraction of this and other carboxylic acids mainly used two neutral extractants, one composed of a mixture of tri-n-butyl phosphate (TBP) [$=C_{12}H_{27}O_4P$] with tri-n-octylamine (TOA) [99, 102], or using mixtures of branched amines [103, 104]. For these purposes, the influence of the chain length of tertiary amines on its extractability has been investigated in polar and nonpolar solvents, respectively. Thus, by using polar solvent such as 1-butanol, 1-hexanol, and 1-octanol, the extractability of tertiary amines shows to increase with respect to the chain length of amines under consideration. In general, the mix of polar and nonpolar solvents has been used in reactive extraction regarding an increase on the efficiency of stripping. Moreover, in many studies where only one kind of tertiary amine such as TOA has been used for the recovery of carboxylic acids not improvements had been observed. Conversely, when a mixture of solvents such as 1-octanol/n-heptane was used in reactive extraction, mixed amine which consisted in relatively short chain tertiary amine and long chain tertiary amine provided higher extractability than that of only one kind of tertiary amine [104]. Furthermore, another study shows that a mixture composed of a linear amine and an acidic one of the organo-phosphoric acid types, the mix induced an important synergic effect in the extraction of carboxylic acids with remarkable biological activity [100]. Pantothenic acid, for example, also known as vitamin B_5, is the amide formed by pantoic acid with β-alanine (Fig. 4.1). This metabolic compound is an infinite water-soluble vitamin involved in the conversion of linear carbohydrates and proteins into sources of energy, like glucose, in the synthesis of coenzyme A, in the synthesis and metabolism of several proteins, and in supporting the body's immune system by strengthening key vitamins [100].

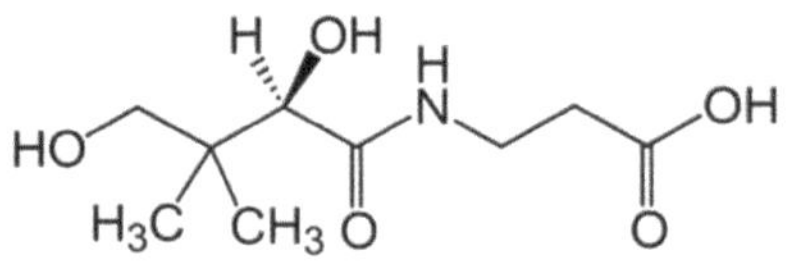

Figure 4.1 Chemical structure of pantothenic acid, vitamin B_5.

In general, pantothenic acid is obtained by extraction methods from natural sources (e.g., bread yeast, cereals, eggs, lentil, mushrooms, peanuts, soya beans, liver of various animals or birds, etc.) or by chemical synthesis [105, 106]. Chemically, pantothenic acid contains both acidic and basic groups; and its reactive extraction with extractants from organo-phosphoric acid and several amine types has been studied with promising results [107].

A careful analysis of pantothenic acid extraction by using single TOA as well as a mix of di(2-ethylhexyl) phosphoric acid (D_2EHPA) with TOA was done to evaluate the influence of the ratio of extractants in the mixture, the solution's pH-value, and the solvent's polarity on the magnitude of the synergic effect.

Table 4.1 Number of amine molecules included in the extracted compound

D_2EHPA concentration	Solvent	Number of TOA molecules (n)
5 g/L	n-Heptane	4.11 (≈4)
	n-Butyl acetate	1.91 (≈2)
	DCM	1.06 (≈1)
20 g/L	n-Heptane	1.91 (≈2)
	n-Butyl acetate	2.05 (≈2)
	DCM	0.92 (≈1)
40 g/L	n-Heptane	2.09 (≈2)
	n-Butyl acetate	1.08 (≈1)
	DCM	1.11 (≈1)

Reproduced from Ref. [107].

To do that, the acid was prepared as an aqueous solution at an initial concentration of 4.56 mM with the extraction carried out in three solvents possessing different dielectric constants, as shown in Table 4.1, namely n-heptane (dielectric constant 1.90 at 25°C), n-butyl acetate (dielectric constant of 5.01 at 25°C), and dichloromethane (dielectric constant 9.08 at 25°C). The considered extractants were TOA and D_2EHPA, and they were used individually in mixtures after being dissolved in the three solvents [107].

4.3.2 Case 2. Ionic Liquids: Bipolar Synergism for Algae Biomass Extraction

Algae (macro, micro) are considered the most important renewable sources as potential feedstock for biodiesel production, and microalgae have shown advantages when compared to crops, animal fat, and organic waste materials, which include high photosynthetic efficiency, less requirement of land for cultivation, and none to minimal competition with the food supply chain.

Lipid extraction is one of the most important processes for microalgae biodiesel production, but the process faces constant limitations such as high energy consumption, solvent's environmental concerns, and low extraction efficiency, which are identified as major bottlenecks for large-scale algal biodiesel production [108, 109]. Lipid extraction mainly includes physical, chemical, and biological methods with the aim of loosening to eventually eliminate the use of hazardous solvents together with the purpose of reducing both lipid extraction time and temperature. In this regard, ionic liquids have the advantages of almost no toxicity, high chemical stability during lipid extraction, and ease of operation in several extractor designs, and they have become one of the most promising lipid extractant solvents for the extraction of microalgae lipids [110]. Using mixtures of ionic liquids with methanol, a group of researchers [111] were able to penetrate the cell matrices of microalgal the *Neochloris oleoabundans* strain to break down the cell walls and recover significant amount of lipids with the aim of testing their effects for lipid extraction with ionic liquids classified as:

1. 1-butyl-3-methylimidazoliumtetrafluoroborate [Bmim][BF_4],
2. 1-butyl-3-methylimidazolium-methylsulfate [Bmim][$MeSO_4$],
3. 1-butyl-3-methylimidazolium-dicyanamide [Bmim][DCN], and
4. 1-butyl-3-methylimidazolium-chloride [Bmim][Cl].

Their findings showed that in the extraction of lipids using these different mixtures of ionic liquids under the presence of methanol in a 1:1 ratio, with reaction conditions set up to occur at 65°C and 2 h, the extraction efficiency follows the increasing order of solvation as follows:

[Bmim] [$MeSO_4$] > [Bmim][BF_4] > [Bmim][DCN] > [Bmim][Cl]

This pattern is consistent with the anionic structure for each ionic mix, which highly correlated with the chemical strength presented for these solvents in their ability to extract lipids from microalgae matrices [111].

The quasi-organic electrolytes formed between these mixtures may possess unique electrical properties in that their constituent solvents may be of much lower dielectric strength than water because the ions of organic electrolytes are often those that comprise room-temperature ionic liquids, which are surprisingly miscible with many different organic solvents (alcohols like) of low dielectric strength. Thus, the strong ionic correlation between these solvents in their solvation activity is thought to be the result of a relatively low dielectric screening of the solvent in large percent, the relative high temperature (65°C) that results in additional properties that can substantially deviate from the Debye–Hückel description and the sudden change of solute charge distribution which forces the solvent to rearrange to a new state of equilibrium due to the new solute charge distribution. In the experiments, the group found that lipid extraction generally increased with a decrease in dipolarity/polarizability conditions, if an increase in the acidity values of the ionic liquid occurred; nonetheless, by using hydrophobic and water-immiscible ionic liquids, it showed very low extraction, while under hydrophilic and water-miscible ionic liquids, the process showed to be very good in terms of efficiency (see Fig. 4.2) [111].

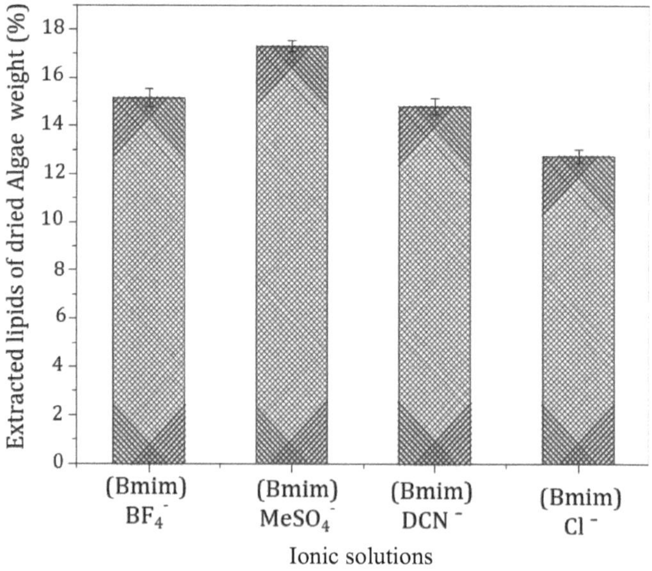

Figure 4.2 Types of ionic liquids and their influence on lipid extraction from *Neochloris oleoabundans*. Reprinted from Ref. [111].

The extraction mechanism promoted by these ionic liquids is thought to occur through the H bonding of the lipids' chemical bulk with the OH groups of molecules in the anionic liquid. It is then proposed that the action of the polar covalent solvent (hot methanol) is largely capable of disrupting the algae's cytomembrane, leading to improvement in the extraction efficiency, and then, lipids are largely extracted from the biomass matrices. Thus, when the cell wall's permeability increases, the excess of methanol molecules in the solvent facilitates the solvation of lipids from algae cells to be dissolved into a bond network that damages cellulose walls that favor the efficiency of extraction of total lipids. The findings also suggest that the solvent formed by [Bmim][Cl]$^-$ and the methanol mixture was not an ideal solution for microalgae lipid extraction because this liquid is more like an absorbent media [112].

The data showed that the lipid extraction efficiency was basically above 15% of algae dry weight with ionic liquids in concentration of 12.5% (1), 25% (2), 33% (3), and 50% (4) (see Fig. 4.3a,b) with the rate gradually decreasing from 20.4% and 21.2 to 6.2% and 5.2%, for mixtures prepared with ionic solutions 2 and 3, respectively.

In addition, the averages of the lipid conversion rate by using each solution (by triplicate) were found to be 16.04%, 15.84%, 14.54%, and 14.59%, in their numerical order of description. Also, the averages of the loss rates for the ionic liquids of 1, 2, and 3 were 7.3%, 6.8%, and 6.4%, respectively.

A significant increase in lipid extraction was observed during the first 2 h with the [Bmim][$MeSO_4$] solution by increasing temperature from 50 °C to 70 °C, and the ionic liquid:methanol solvent in the 1:3 ratio. Under these setup conditions, the extraction efficiency was improved to 17% from 13.5% which occurred with the ionic liquid: methanol solvent in 1:7 ratio is observed in Figs. 4.4 and 4.5, respectively [111].

Using an acidic hydrolyzed algal slurry of *Nannochloropsis* spp. and *Scenedesmus obliquus* microalgae, hexane was used as the solvent for lipid extraction to investigate the effect of mixed solvents for internal mass transfer on the recovery of lipids [113].

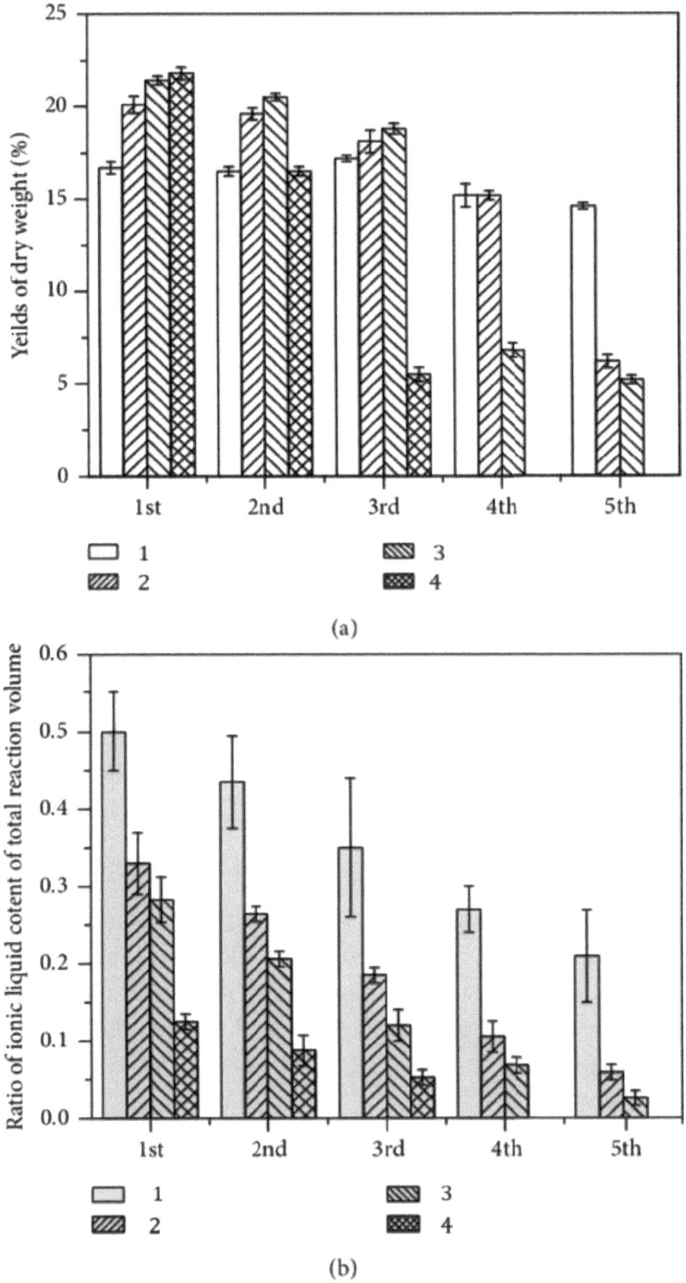

Figure 4.3 Influence of recycle times of [Bmim][$MeSO_4$] on lipid extraction and ionic liquid loss. Reprinted from Ref. [111].

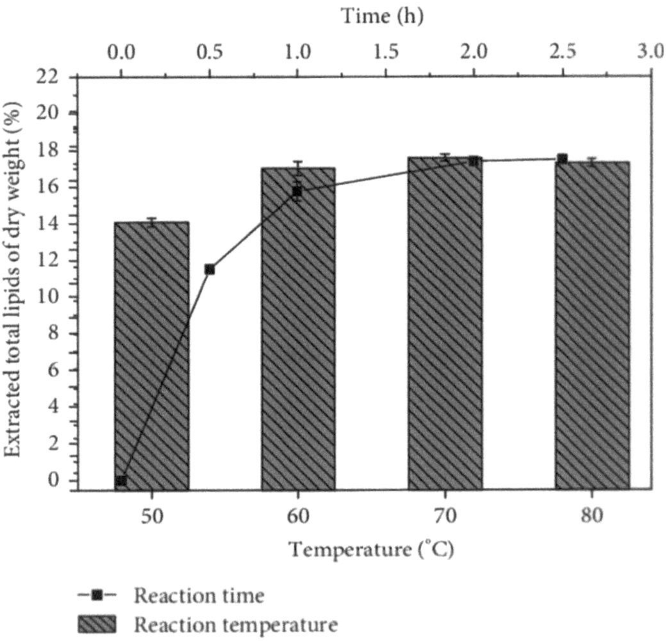

Figure 4.4 Effects of temperature and time on lipid extraction from microalgae biomass through ionic liquids and methanol synergistic activity. Reprinted from Ref. [111].

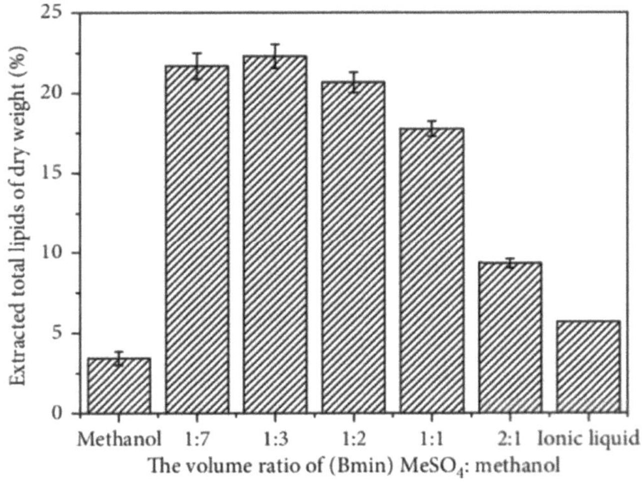

Figure 4.5 Effects of volume ratio of [Bmim][$MeSO_4$] : MeOH ionic solution on lipid extraction (wt. %) from microalgae. Reprinted from Ref. [111].

4.4 Role of Synergism in Extraction Technologies

To mitigate the escalated deterioration of our shared environment, the research community working with extraction technologies has been suggesting that with the increased number of solvent-assisted extraction alternatives, it could be more safe by using low to nonvolatile organic solvents for extraction or partition separation processes that lead to a chemical analysis, as these types of solvents possess less to no risks like the minimum advantages listed in Table 4.2. In this proposal, however, the short list makes special reference to the analytical advantages associated with the use of supercritical fluid extraction (SFE), one of the several recent alternative methods proposed to address one of the chemical limitations related with organic solvents due to their association with health and environmental concerns. Herein, we intend to present the most fundamental descriptions associated with upgraded techniques recently designed to eliminate or substantially reduce the dependence of organic solvents in extraction processes performed in academia, biotechnology, and laboratory environmental analysis with some of them understood as methods that work best under synergistic accomplishment.

Table 4.2 Current techniques used as synergistic or complementary methods for solvent extraction

1.	Microwave extraction
2.	Pulsed electric field extraction
3.	Pressurized fluid extraction
4.	Supercritical fluids, particularly supercritical carbon dioxide (SC-CO_2)[a]
5.	Using membranes for extraction[b]
6.	Immunoassay-based methods[c]
7.	Miniaturization of Soxhlet extraction, reagent chemical analysis[d]
8.	Pythonics extraction
9.	Solid phase microfiber extraction[e]
10.	Solid phase extraction[f]

[a]Discussed in Chapter 5 [114], [b][115], [c][116], [d][152], [e][153], [f][154, 155]

4.4.1 Microwave, Ultrasound, and Pulsed Electric Field Extraction [MW-US-PEF]

The use of ultrasound-assisted waves on the extraction of lipids has recently been widely accepted as an alternative method, which is expected to overcome the difficulties encountered with the use of conventional mechanical disruption methods (e.g., pressing).

The MW-US-PFE methods are carried out in a non-sophisticated equipment under easy working conditions by imparting higher purity to the final product and eliminating costly treatment of wastewater that is generated during the pressing process. Furthermore, these techniques are shown to be more economical and eco-friendly because the process can be completed in a very short time with high reproducibility when compared with traditional extraction technologies [117].

An interesting benefit is the very little need for energy input when compared to that necessary to carry out short experiments or for the long time in continuous regimes carried out on conventional methods but requiring to be operated at lower temperatures. When applied on liquid cell cultures, there are two major mechanisms by which continuous ultrasonic or pulsed waves can cause damage to cell walls, namely, cavitation and acoustic streaming, respectively.

- *Cavitation* causes the production of microbubbles inside the cell matrices because of specific applied ultrasound frequencies, which in turn can create internal pressure in the space's matrix, which breaks up the walls following the wave's direction [117].
- *Acoustic streaming*, on the other hand, facilitates the mixing of solute/solvent molecules in the extracted material [118].

Thus, under a stable and controlled voltage application, ultrasonic waves, also described as pulsed electric field (PEF) method, will generate transient and constant cavitation, which produces compression/decompression cycles in the cell matrix, which results in an increase in the matrices' internal temperature, which in turn will facilitate the solubility of most thermolabile light molecules, as shown in Fig. 4.6 [119].

The activation of labile molecules and the heat generated in this process may weaken or break the cell walls so that the bioactive compounds can be released easily from their matrices as material more available to dissolve in the extraction solvent(s). Unsteady wave oscillations will result in transient cavitation, which will ultimately implode because of rapid changes in the matrix's internal pressure. When constant cavitation is applied to organic tissues or vegetal cell matrices, it will result in an implosion, which produces extremely localized heat shock waves, which disrupt the cellular cohesion and lead to matrices breaking down [120]. The physical effects of wave sonication are cracks on the cell wall or membrane rupture due to the cavitation effects on the liquid content, which is found inside every cell matrix. Hence, microstreaming and heightened mass transfer are the result of cavitation together with bubbles collapsing as two main critical steps that determine the lipid yield extraction efficiency by using the microwave extraction method [119].

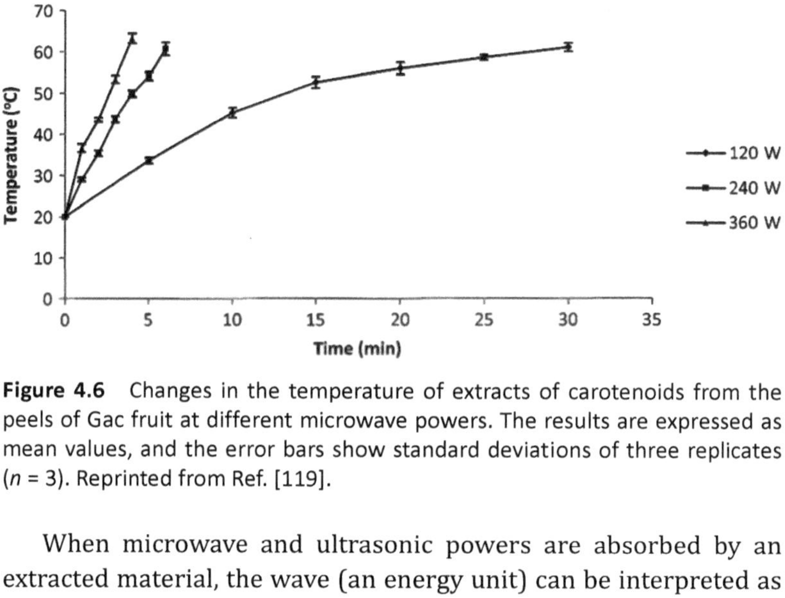

Figure 4.6 Changes in the temperature of extracts of carotenoids from the peels of Gac fruit at different microwave powers. The results are expressed as mean values, and the error bars show standard deviations of three replicates ($n = 3$). Reprinted from Ref. [119].

When microwave and ultrasonic powers are absorbed by an extracted material, the wave (an energy unit) can be interpreted as the mass unit of the extraction solution to be determined through the equation:

$$P = C_p \frac{dT}{t} \tag{4.2}$$

where P is the power absorbed by a mass unit of the extraction solution (W/g), C_p is the specific heat capacity (J/g°C), ΔT is the temperature change by the cavitation process (°C), and t is the experiment's time (s).

Thus, the controlled application of pulsed electric fields (PEF) during specific periods makes this an efficient technique for the extraction of bio-food and other agricultural bioproducts to be integrated as a popular and widely used extraction alternative in recent times [121]. The technique had demonstrated that by using short duration pulses (10^{-4} to 10^{-2} s) at scan rates of 500–1000 V/cm, this type of physical conditions will affect substantially the cell wall properties for its permeation, which characterize biological membrane barriers by allowing an accelerated activity of solvents during the extraction by facilitating the solute's mass transport during the process. In addition, the technique avoids other physical undesirable changes on the extracted biological material, which are part of experimental limitations observed with other techniques, such as thermal degradation, solute chemical composition, and enzymatic oxireduction [122] within others. A supplementary advantage of the pulsed electric field technique, specially when used for food extraction applications, is related with the possibility of its activity on microbial and pathogen microorganism killing [123]. This technology shows the potential to improve energy usage economically and efficiently, besides the advantage of providing microbiologically safe and minimize contamination of processed foods. The successful application of pulsed electric field as extraction technology suggests it can be an alternative substitute for conventional thermal processing of liquid food products such as fruit juices, milk, liquid egg, vegetable oils, and supplement food [123]. In general, a PEF system for food processing consists of three basic components (as shown in Fig. 4.7): (1) a high-voltage pulse generator, (2) a treatment chamber, and (3) an automated system that takes control of all process parameters [124].

At this point, however, it is important to highlight that PEF is understood mainly as an assisted technique that displays unusual synergetic physical effects with several possibilities to be processed at controlled temperatures.

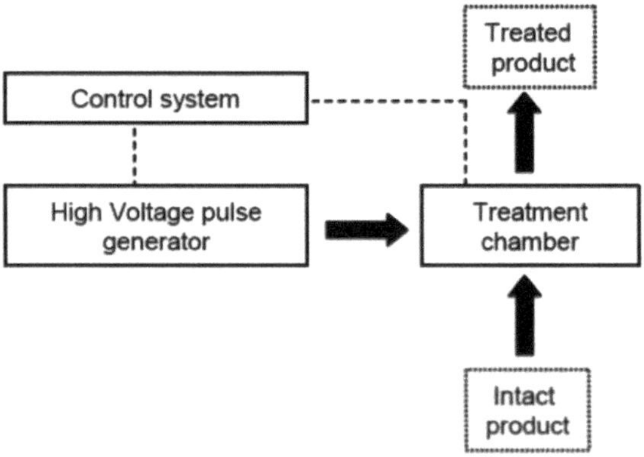

Figure 4.7 Scheme of a pulsed electric field system for food processing. Reprinted from Ref. [124].

4.4.1.1 The treatment chamber in pulsed electric field

During a PEF extraction, one of the most important and complicated components in the design system is related with the treatment chamber. This key component keeps the treated product inside the extractor during the steps of pulsing, although the compact uniformity of the process is highly dependent on the design of the treatment chamber. Thus, if the strength of an applied electric field exceeds the electric field supported for the food product treated in the chamber, an easy breakdown of food particulate occurs as a spark. Therefore, treatment chambers are mainly designed to be operated either in a batch of containers or in a continuous manner; thus, batch systems are generally found in early designs for handling of static volumes of solid or semi-solid foods. In these cases, several treatment chambers must be designed and categorized mainly into two types: (1) parallel plate and (2) coaxial plate, as observed in Fig. 4.8. Thus, parallel-plate chambers have been typically used in batch process, whereas coaxial plate designs are mostly used in continuous extraction with the medium pumped through a known flow rate and electrical pulses also applied at a known pulsed frequency.

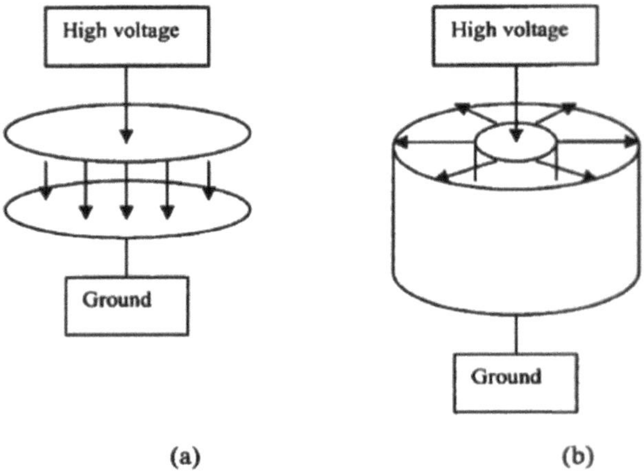

Figure 4.8 Diagrams of common electrode configurations for pulsed electric field treatment chambers: (a) parallel plate and (b) coaxial plate. Adapted from Ref. [124].

On the other hand, coaxial chambers, when used in continuous regimes, have been found to result in higher inactivation rates compared to batch systems since there is more uniform distribution of the electric field in continuously flowing media, as shown in Fig. 4.9, which presents different chamber designs.

Under any circumstance, an electrolytic solution may be used to facilitate the electrical conductivity between the electrodes, and/or an ionic semi-permeable to permeable membrane can be used to satisfy this physical chemistry requirement. Thus, to set up an extraction system, suitable electrolyte solutions can include sodium carbonate, sodium hydroxide, potassium carbonate, potassium hydroxide, and weak acid solutions (e.g., acetic acid). These solutions are circulated continuously to remove the products of electrolysis and be replaced in the event of their excess concentration or depletion of ionic components, which leads to plugging narrow conducts or undesirable by-products deposition.

The successful application of pulsed electric field as an extraction technology has been demonstrated in, for example, the pasteurization of foods such as juices, milk, yogurt, soups, and liquid eggs (see discussion in Chapter 6).

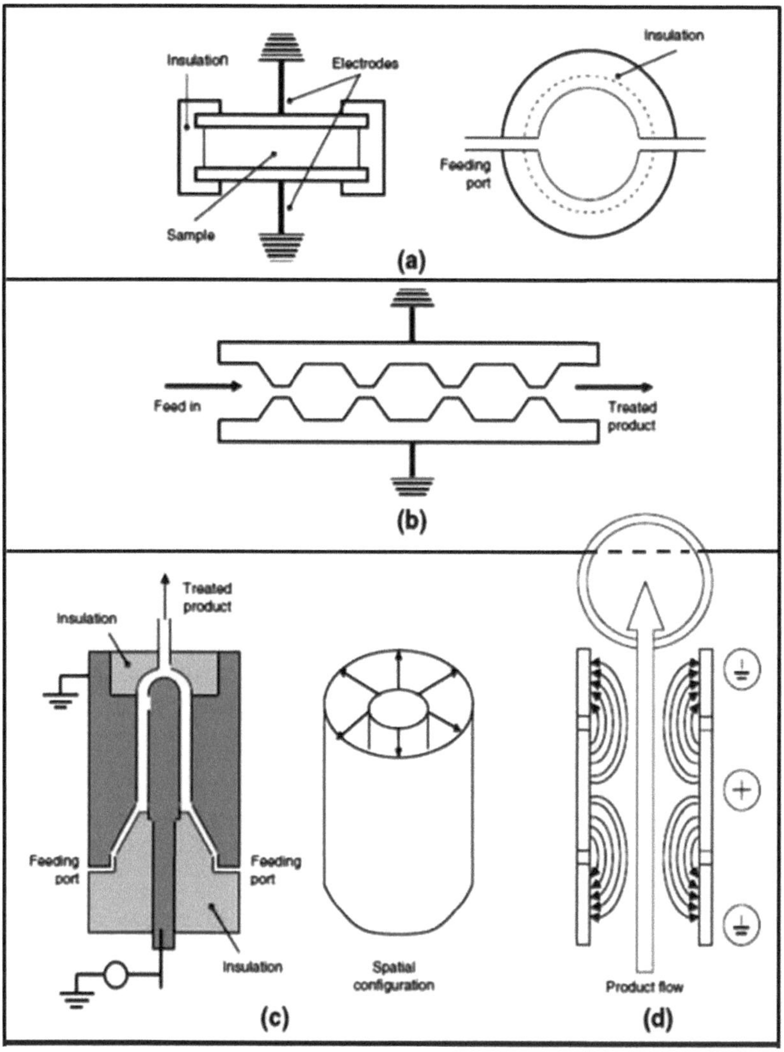

Figure 4.9 Schematic diagram of a pulsed electric field system with treatment chambers: (a) static chamber, (b) side view of a basic continuous design, (c) a coaxial chamber, and (d) a colinear chamber. Adapted from Ref. [124].

However, the PEF technology has been mainly applied for liquid food products conditioned to avoid the presence of air bubbles during the process among be limited to operate in the form of short pulses to avoid excessive heating on valuable molecules with heat sensitivity for degradation or undesirable electrolytic reactions with aims to prevent corrosion effects on electrodes surface. In addition, the maximum particle size dispersed in the liquid must be smaller than the gap of the treatment region in the chamber to ensure the best results. Therefore, PEF technology need to be processed as a continuous method by doing of it a limited and not at all suitable to be applied for samples that are not pumpable food products.

Nonetheless, PEF methods can be applied to enhance the extraction of organic sugars and other cellular content from plant matrices, such as sugar beets, flavonoids, and antioxidants from fruits or dry crops, and the method could also find application in reducing the solid volume of wastewater (sludge suspension) throughout oxireduction reactions.

4.4.1.2 Advantages/disadvantages of ultrasound-pulsed waves for natural compounds extraction

Advantages

- Assisted pulsed electric field extraction is thought to be an inexpensive, simple, and efficient alternative when compared with conventional extraction techniques.
- It is credited that the main benefits encountered using pulsed waves for a solid in a liquid extraction process include a substantial increase in product yield because of the method's faster kinetics.
- Ultrasound waves operating during short time spaces also reduce the process temperature, a parameter determinant in the chemical stability of thermolabile molecules [125].
- When compared with other novel extraction techniques such as microwave-assisted extraction, an ultrasound apparatus is cheaper, and its operation is easier too.
- Ultrasound-assisted extraction and a Soxhlet extractor system can be used in a synergistic approach by employing any solvent for the extraction of a wide variety of natural and raw materials [125].

Disadvantages

- A known deleterious effect of ultrasound wave energy (plus 20 kHz) relays with the active constituents of medicinal plants through the formation of free radicals and consequently undesirable changes in the drug molecules [126].

- Food products need to avoid the presence of air bubbles.
- Food and animal products need to be handled under low electrical conductivity voltages to prevent electrochemical reactions.

4.4.2 Pressurized Fluid Extraction

Pressurized fluid extraction (PFE) is a US-EPA regulated method identified with the SW-846 code and is preferably recommended for the extraction of hazardous heavy semi-volatile organic compounds, organophosphorus and organochloride pesticides, chlorinated herbicides, polychlorinated biphenyls (PCBs), polychlorinated dibenzo-p-dioxins (PCDDs), polychlorinated dibenzofurans (PCDFs), and diesel range organics (DRO), which may then be analyzed by a variety of chromatographic technologies [127].

Furthermore, the method is an analytical procedure for the extraction of water-insoluble or slightly water-soluble organic compounds from matrices like soils, clays, and from deposits such as sediments, sludges, and waste solids. It is highly recommended that the method may be carried out at elevated physical parameters of temperature and pressure (over 100°C and >1500 psi) with the aim to achieve analyte recoveries equivalent to those obtained by Soxhlet extraction but using less volume of solvent and significantly decreasing the time necessary to achieve same yields as when using Soxhlet methods.

The high pressure and temperature necessary to proceed with an extraction make the method require the use of a robust equipment that necessarily needs to be constructed in stainless steel materials that fulfill safety regulations and guarantee the equipment will support extreme physical conditions.

In addition to analytical environmental determination, when PLE is applied for the extraction of natural materials, the technique has also been referred to as accelerated solvent extraction (ASE), pressurized fluid extraction (PFE), pressurized hot solvent extraction (PHSE), high-pressure solvent extraction (HPSE), high-pressure high-temperature solvent extraction (HPHTSE), as well as subcritical solvent extraction (SSE) [128].

Independent of any technological definition, this extraction technique allows the employment of solvents at both high temperature and pressure as mentioned above, but always below their respective critical points. Therefore, the solvent is maintained in the liquid state during the whole extraction procedure by given it remarkable advantages.

As a direct result of utilizing non-critical conditions related with temperature and pressure, mild changes in the solvent physicochemical properties must occur in favor of the process. For instance, the mass transfer rates are enhanced in the extract solution, and at the same time, both surface tension and viscosity of solvent are decreased and, then, an increase in the analyte's solubility is highly favored. Through these thermodynamic activities in place, the solvent can act quickly and penetrate easier and deeper into the solid matrix to dissolve molecules targeted to be extracted. Thus, it is expected that working under continuous extraction conditions, high extraction yields are obtained when compared with results using a conventional extraction apparatus.

In terms of efficiency and economic impact, not only the use of PLE may result in a faster extraction process, but also this lowered the solvent consumption for sample preparation of solids. In addition to these technological advantages, most of the instruments used for PLE are automated and allow the development of less labor-intensive extraction by improving reproducibility and efficiency (Fig. 4.10).

Solvents other than water have been explored for essential oils extraction. In this regard, ethanol, ethyl acetate, and hexane have been recently evaluated as efficient solvents for the extraction of α-bisabolol-rich oil from *Eremanthus erythropappus* wood, using PLE and UAE techniques [129]. The importance of α-bisabolol extraction is its pharmacological content of essential oils, which are found in several plant genus and are widely used in dermatologic formulations, decorative cosmetics, fine fragrances, shampoos, among others. When PLE is done at mild conditions (55°C), a high purity (64.23%) in terms of α-bisabolol content was obtained using n-hexane, while UAE provided the highest yield using the same solvent.

4.4.2.1 Pressurized liquid extraction for antibiotics and pharmaceuticals as soil–water contaminants

The presence of pharmaceutical products in any one of the three environment compartments (air, soil, and water) has raised health issues concerning their occurrence and effects on the biota elements.

Antibiotics are an important group of pharmaceutical products widely used in human and animal health care, which are reported to be ubiquitous compounds in the aquatic and soil environments.

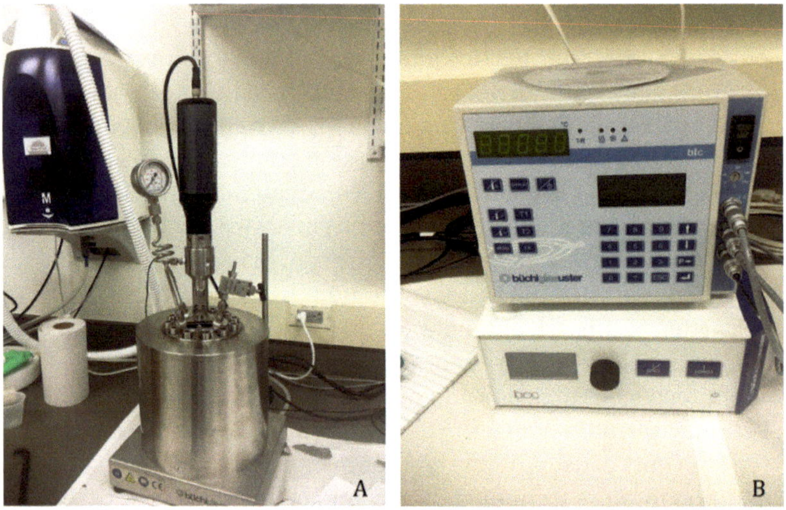

Figure 4.10 Jacketed and mechanical stirred calorimeter Cyclone-075 Büchiglasuster model (A) with its control panel (B) (speed manually adjustable 0–3000 rpm) for automated extraction. The stirred reactor is in Hastelloy steel, and it has a nominal volume of 250 mL. It can resist up to 250°C and 60 bars of internal pressure. Picture from the author's lab.

As discussed initially for other techniques, PLE is also thought to be a reliable method for the extraction of antibiotics and other drug residues associated with suspended solid matter [127]. Thus, the analysis of multiresidue from sulfonamide antibiotics and their acetylated metabolites as contaminants of soils and sewage sludge (sediments) that can be performed using a fully automated PLE system (Fig. 4.10) with the extracts analyzed by liquid chromatography coupled to an electrospray-quadrupole linear ion trap mass spectrometer (LC-QqLIT-MS) [130].

To ensure the completion of analysis, a subsequent solid phase extraction (SPE) (the synergy step) for preconcentration and purification is required, by taking into consideration the complexity of handle sludge samples by using, for instance, a hydrophilic–lipophilic balanced polymer membrane.

For specific considerations, the PLE methodology was applied to evaluate the occurrence of the target sulfonamides in several sewage sludge matrices and soil samples collected in different wastewater treatment plants and agricultural areas [130]. The process starts with an assisted solvent extraction of quinolone and sulfonamide residues (e.g., lomefloxacin, enoxacin, sarafloxacin, enrofloxacin, sulfadiazine, sulfamethoxydiazine, and sulfadimethoxy-pyrimidine) in fish (e.g., sardine) and shrimp, carried out by PLE using a dispersant agent like diatomaceous earth and acetonitrile as extraction solvent [131]. The process rendered very good recovery, and total efficiencies were found to be 60–130% for most of the sulfonamides in both matrices at two spike levels.

The final report indicates that through PLE and SPE methods, it was possible to confirm the presence of sulfonamides in soil and water matrices, with sulfathiazole and sulfamethazine as the two sulfonamides most frequently detected in the analyzed sewage sludge and soil samples, respectively [131].

4.4.2.2 Pressurized liquid extraction of fatty acids for vegan and nutraceutical products

The extraction of valuable nutrients from algae biomass for its subsequent use as feedstock in specialized food ingredients has widely increased its consumer acceptance. Over the past few decades, many medicinal and bioactive properties related with algae biomass have been investigated and several algae extracts have been brought to the market as nutritional food supplements. Some examples include Solaray®, a dietary supplement product that contains the extract of the red marine algae *Rhodymenia palmata*, which is credited to maintain a good response of the immune system; Omega 3 Vegan Lineavi, a product derived from *Schizochytrium* sp. algae oil, which provides a vegetarian and vegan friendly source of healthy fatty acids [132]. Thus, Omega-3 (ω-3) fatty acids (FAs) are mainly found in unicellular phytoplankton and several marine algae, which are natural food of the marine biota. When the biota was analyzed, eicosapentanoic acid (EPA) was found, which accumulates in all those marine species that consume algae and get passed to other species through the food chain [35]. These oil nutrients have been found to have positive effects on the human central nervous system

(HCNS) for healthier brain function, retinal and neural tissues in fetuses, and growth of young children [133].

Those and other socioeconomic benefits have generated great interest in the biotechnology of algae biomass extraction as a source of products containing highly valuable FAs, intended to be used in nutrition and wellness consumption; however, advanced methods for maximum extraction and best analytical determination of different FAs contained in algae matrices are highly needed. Several solvents have been used to extract, separate, and purify algae biomass with the most common solvents being n-hexane, chloroform, dichloromethane, acetone, methanol, and mixtures of hexane: acetone [134], chloroform: methanol [135] in different proportions.

Together with the variety of solvents recommended, several methods have been widely recommended to extract lipids and bioproducts from natural sources (algae being the most popularly used): solid–liquid extraction (SLE), ultrasonic solvent extraction (USE), and supercritical fluid extraction (SFE). And PLE is currently considered an advanced technique because it offers important benefits such as shorter extraction time, low solvent consumption, low sample handling, and increased product yield, in addition to be considered a technique aligned with the green aspects of sample preparation [136].

In an initial case study, the lipid contents of four macroalgae Phaeophyta species—(1) *Fucus vesiculosus* (*F. vesiculosus*), (2) *Dictyota dichotoma* (*D. dichotoma*), (3) *Cystoseira baccata* (*C. baccata*), and (4) *Himanthalia elongata* (*H. elongata*)—and two Chlorophyta species—(1) *Ulva intestinalis* (*U. intestinalis*) and (2) *Ulva lactuca* (*U. lactuca*)—were compared.

Every algae sample was treated as an ethanolic solution to be extracted in a PL extractor with the six extracts evaluated for their antioxidant and antibacterial activity, respectively [136].

4.4.2.3 Effects of pressurized liquid extraction on macroalgae oil lipid composition

In other study, pressurized liquid extraction was used as an unknown assessment to evaluate the effect this extraction method has on lipid chemical composition of macroalgae biomass that until the time of the proposed study, non-assessment was reported on this issue so far. Thus, several studies on optimized PLE had been published in the extraction of carotenoids and phenolic compounds from microalgae (see Section 4.2); however, nobody had reported about the use of PLE technique to extract FAs from macroalgae with the same interest.

According to the published assessment, *F. vesiculosus* strain accumulate many fatty acids so far, highlighting on the effect of lipid profile extraction through different PLE conditions and how it could be investigated by other researchers using this algae as a matter of proof of these claims [137]. Therefore, two physical parameters that show a significant effect on PLE recovery efficiencies were investigated, which include the solvent's type and temperature.

The experiments were carried out on 1.0 g of grinded *F. vesiculosus* macroalgae and submitted to PLE by using four pure solvents and a polar mixture (acetone, hexane, ethanol, ethyl acetate, and ethanol: water mix 50:50 - v/v) at three temperatures (80°C, 120°C, and 160°C) [138].

These selected solvents possess a wide range of dielectric constants and, then, are able to extract bioproducts possessing different polarities too. These polar particularities could offer advantages in the choice of solvent for purification and separation when using the partitioning method (see Chapter 2). The extraction was set up to complete after 10 min, with the resulting extracts collected, dried, and weighted to determine the product yield, which are presented in Table 4.4.

By comparison, the highest yields related with PLE activity were obtained with solvent(s) showing the most polar strength (ethanol: water 50:50, ethanol, and acetone), while the lowest yields were obtained with weak polarity solvents such as ethyl acetate and hexane, respectively. Thus, at 120°C (the intermediate temperature), 414.9 ± 14.7 mg was obtained using ethanol: water 50:50 solution, 119.8 ± 11.8 mg with ethanol, 107.3 ± 2.3 mg with acetone, 55.9 ± 2.1 mg with ethyl acetate, and 37.2 ± 2.4 with hexane.

It has also been observed that the lipid yield in this algae strain is dependent on the extraction temperature applied in the PLE system, reaching the maximum amount at 160°C in the ethanol: water 50:50 solution (571.9 ± 23.8 mg) [138].

The lipid contents were evaluated by gas chromatography coupled with a mass spectrometer (GC-MS) technique.

Table 4.3 Lipid content (%) and FA (mg/g) composition of six algae species (DW)

	FA (C:U)	RT (min)	*F. vesiculosus*	*C. baccata*	*H. elongata*	*D. dichotoma*	*U. lactuca*	*U. intestinalis*
	FA 14:0	9.802	11.09 ± 0.19	5.13 ± 0.12	1.72 ± 0.04	3.01 ± 0.37	1.78 ± 0.22	1.96 ± 0.23
	FA 15:0	10.774	0.31 ± 0.03	N.D	0.17 ± 0.03	N.D	0.18 ± 0.02	0.08 ± 0.00
	FA 16:1	11.789	0.98 ± 0.22	2.19 ± 0.04	0.56 ± 0.01	1.16 ± 0.05	0.16 ± 0.00	0.16 ± 0.00
FA content	**FA 16:0**	12.078	9.64 ± 0.30	6.80 ± 0.29	5.85 ± 0.14	4.40 ± 0.64	6.09 ± 0.29	6.02 ± 0.22
(mg/g algae)	**FA 18:3**	14.900	0.08 ± 0.00	N.D 0.16 ±	0.04 ± 0.04	N.D 0.01 ±	0.09 ± 0.01	N.D 0.06 ± 0.00
	FA 18:2	15.304	0.34 ± 0.04	0.02	0.01 ± 0.00	0.00	0.05 ± 0.01	
	FA 18:1	15.507	13.15 ± 1.03	3.09 ± 0.34	0.49 ± 0.09	1.09±0.05	0.47 ± 0.02	0.23 ± 0.01
	FA 18:0	16.041	1.56 ± 0.13	1.65 ± 0.16	1.80 ± 0.04	1.28 ± 0.07	1.68 ± 0.11	2.11 ± 0.08
	FA 20:4	20.549	1.30 ± 0.12	0.62 ± 0.01	N.D	N.D	N.D	N.D
	FA 20:5	20.806	0.36 ± 0.08	0.24 ± 0.01	N.D	0.15 ± 0.03	N.D	N.D
FA total (mg/g algae)	38.83	19.87	10.64	11.09	10.46	10.63		
Lipid content by Folch (%)	6.6%	6.7%	6.0%	5.7%	4.8%	4.6%		

Lipids were extracted using the Folch method ($n = 3$).
DW = dry weight; number of carbon and unsaturation (C:U) status and retention time (RT) of the fatty acid methyl esters (FAMEs) are also included. Results show mean standard error of the mean (SEM) of three experiments. N.D means not detected. Adapted from Ref. [138].

Table 4.4 Yields (%) of lipids obtained from macroalgae *F. vesiculosus* through PLE

Algae strain: Fucus vesiculosus					
Yield (%) of lipids by PLE					
Extract temp.	**Hexane**	**Ethyl acetate**	**Acetone**	**Ethanol**	**Ethanol: water 50:50**
80°C	2.79 ± 0.12	4.72 ± 0.11	9.01 ± 1.24	8.85 ± 1.51	34.85 ± 3.11
120°C	3.72 ± 0.24	5.59 ± 0.21	10.73 ± 0.23	11.98 ± 1.18	41.49 ± 1.47
160°C	4.49 ± 1.54	7.03 ± 1.79	12.90 ± 1.21	12.89 ± 0.68	57.19 ± 2.38

Results show mean ± standard error of the mean of three experiments. Data reprinted from Ref. [138].

From the data collected through this separation technique, it was possible to identify up to 11 FAs in the macroalage biomass, which includes 10 previously detected lipids in an extraction performed by the Folch method and eicosa-5,8,11-trienoic acid (C 20:3) as small quantities at the end of the chromatogram (RT = 21.233).

These data, in particular, are very important because they relay with the health of human tissues as this acid (C 20:3-9) is a physiological regulator of the amount of arachidonic acid in tissues and is commonly denoted as "omega 3 or 6," with the n-3 and n-6 polyunsaturated fatty acids (n-3 and n-6, PUFAs), which are constituents of human diet and found both in vegetables and animal tissues [139–141].

Thus, in the range of temperature used for FAs extraction the profiles showed to be almost similar, according with the data reported in Table 4.5. The data shows that lipids composition were not substantially affected by either temperature, nonetheless the lower yields were obtained at 80 °C as observed by the low intensity of the GC peaks followed for those carried out at 160 °C with the high signal obtained at 120 °C, therefore, it was selected the mean temperature of 120 °C as the optimum experimental parameter to obtain *F. vesiculosus* extracts presenting the high lipid yields [138].

Table 4.5 FA composition (mg/g) of lipids extracted by PLE from *F. vesiculosus* macroalgae with five solvents (acetone, ethanol, hexane, ethyl acetate, and ethanol: water 50:50) at 120°C

	Algae strain: Fucus vesiculosus				
	FA quantity (mg/g PLE)				
FA	**Hexane**	**Ethyl acetate**	**Acetone**	**Ethanol**	**Ethanol:Water 50:50**
FA 14:0	150.51 ± 9.83	227.31 ± 9.43	247.35 ± 5.07	211.62 ± 5.19	29.36 ± 0.52
FA 15:0	2.53 ± 0.15	3.8 ± 0.42	2.97 ± 0.25	2.91 ± 0.11	0.12 ± 0.00
FA 16:1	10.67 ± 0.27	14.21 ± 0.60	14.78 ± 0.56	13.33 ± 0.5	1.89 ± 0.05
FA 16:0	91.37 ± 8.55	150.37 ± 6.01	167.51 ± 7.42	159.16 ± 14.68	84.98 ± 2.04
FA 18:3	0.16 ± 0.27	0.71 ± 0.12	0.39 ± 0.01	0.46 ± 0.07	N.D
FA 18:2	3.76 ± 0.32	6.18 ± 0.17	3.58 ± 0.02	3.47 ± 0.64	0.45 ± 0.02
FA 18:1	131.52 ± 10.54	234.44 ± 6.62	123.19 ± 28.83	126.11 ± 8.48	18.33 ± 0.32
FA 18:0	12.22 ± 1.46	19.75 ± 0.77	16.19 ± 1.21	16.92 ± 1.62	16.43 ± 0.22
FA 20:4	15.65 ± 1.49	23.38 ± 0.82	11.41 ± 0.01	13.02 ± 2.29	2.92 ± 0.04
FA 20:5	6.63 ± 0.66	11.37 ± 0.39	6.94 ± 0.47	7.68 ± 0.58	1.75 ± 0.02
FA 20:3	1.10 ± 0.10	1.68 ± 0.30	0.87 ± 0.05	0.91 ± 0.97	N.D
Total FA	426.12	693.20	595.90	554.42	156.23
Total ω-3	6.63	11.37	6.94	7.68	1.75
Total ω-6	19.57	30.27	15.38	17.43	3.37
Ratio Total ω-6/Total ω-3	2.95	2.665	2.215	2.208	1.92

Results show mean ± standard error of the mean of three experiments. N.D means not detected. Reprinted from Ref. [138].

The data presented in Table 4.5 correspond to analytical findings obtained through gas chromatography (GC) and are observed in Fig. 4.11 with extracts collected at 120°C on hexane (A), ethyl acetate (B), acetone (C), ethanol (D), and ethanol: water 50: 50 (E), respectively.

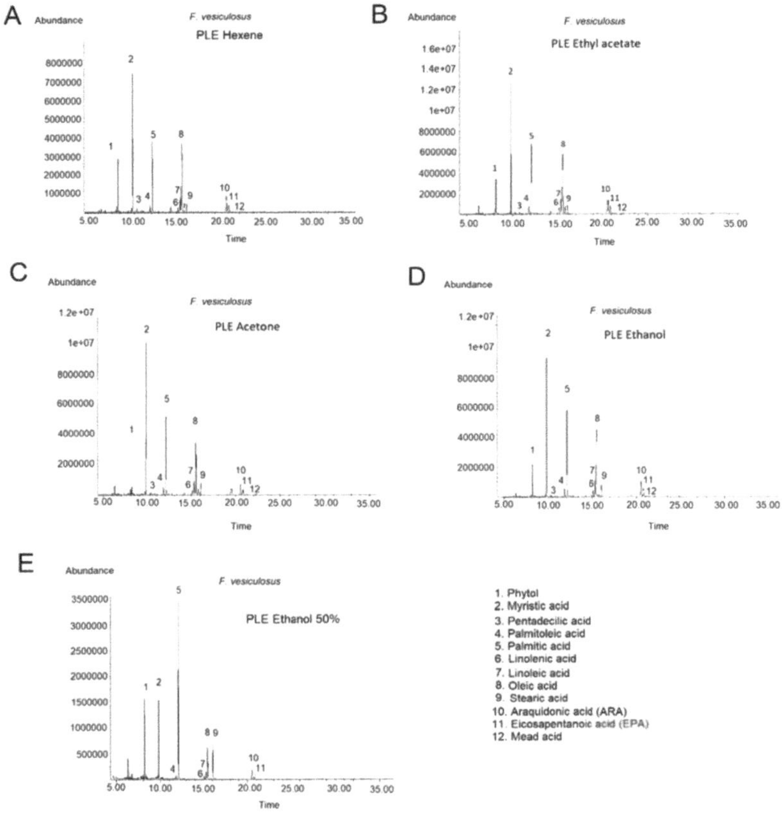

Figure 4.11 GC-MS chromatograms of *F. vesiculosus* macroalgae extracted at 120°C through PLE with hexane (A), ethyl acetate (B), acetone (C), ethanol (D), and ethanol: water 50: 50 (E). The column used for FA separation was an Agilent HP-5MS UI capillary column (30 m × 0.250 mm × 0.25 μm) with oven temperature programmed to start at 50°C, increased to 210°C at 20°C per minute and hold for 18 min. To afford the whole separation, the temperature was further increased to 230°C at 20°C per minute and kept at 230°C for 13 min. Reprinted from Ref. [138].

The GC showing the maximum FA content was with *F. vesiculosus* extracted with ethyl acetate (693.20 mg), followed by acetone (595.90 mg), and ethanol (554.42 mg) as total FA/g PLE.

Conversely for a synergy process, ethanol:water 50:50 (156.23 mg) was not the best choice for FA extraction from this alga, but interestingly, the use of hexane (426.12 mg) did not improve the FA extraction with regard to the use of acetone or ethanol individually. Therefore, it was concluded that the extraction profiles of lipids obtained with solvents possessing different dipole moment, that their yield proportion were pretty much different and dependant on the solvents' polarity according with the data presented in Table 4.5 and Fig. 4.11, respectively.

The data show that extraction carried out with ethyl acetate enhanced oleic acid (C 18:1, peak 8), araquidonic acid (ARA) (C 20:4, peak 10), and eicosapentanoic acid (EPA) (C 20:5, peak 11) yield, while in experiments carried out with acetone and ethanol, these quantities were almost double. In contrast, the use of ethyl acetate did not enhance the extractions for myristic (C 14:0, peak 2) and palmitic (C 16:0, peak 5) acids (Fig. 4.11b).

Thus, the compounds that were enhanced are those with chemical structure characterized by long chains of FAs because they are less polar and their solubility in polar solvents is strongly limited when compared with the molecules of short chains of fatty acids.

With the same purpose, other authors have used supercritical liquid extraction (SLE) and/or PLE with solvents of different polarity for the extraction of FAs from food and microalgae matrices. For example, Castejón et al. did not observe significant changes in the chemical composition of *chia seed* oils extracted with solvents possessing different polarity [141]. Also, da Costa Rodriguez et al. studied the effect of the solvent's chemical composition on the extraction of rice bran oil extracted with ethanol containing different water percent in the range of 60–90°C. From their studies, they concluded that there were great differences in the lipid's composition obtained at different temperatures with the extraction yield being highly affected by the solvent water content, and it varied from 8.56 to 20.05 g of oil/100 g of fresh rice bran (or 42.7–99.9% of the total oil available) depending on the experimental conditions [142].

4.4.2.4 Phytonics extraction: the synergism of fluorocarbon solvents

In search for new solvent-assisted extraction opportunities, a solvent-based technology termed "hydrofluorocarbon-134a" has emerged as a novel method to optimize some remarkable chemical properties of a solvent for its use in the extraction of plant material by offering significant environmental advantages on its physical chemistry and safety benefits over traditional processes for the extraction of high-quality natural fragrant oils, flavors, and specific biological extracts.

The method had been implemented and described by Advanced Phytonics Limited (Manchester, UK) enterprise, which has developed this patented technology termed "phytonics process" to be performed in an equipment, as proposed in Fig. 4.12 [143].

Figure 4.12 Picture of a 50 L Advanced Phytonics Extraction Unit. Source:https://rccostello.com/wordpress/extraction/advantages-advanced-phytonics-extraction-systems/.

The products extracted throughout the process are mainly classified as fragrant molecules of essential oils in addition to biological and phytopharmacological extracts, which are described to be used directly without further physical or chemical treatment.

The efficiency of the method is based on the physical advantages and chemical properties exhibited for a new generation of fluorocarbon solvents, which can be used for the extraction of several plant materials, which have wide medicinal applications [143].

It should be highlighted at this point that the core of this chemical technology rests on a solvent system with the formulae 1,1,2,2-tetrafluoroethane, better known as hydrofluorocarbon-134a (HFC-134a) developed with the purpose of replacing the use of hazardous chlorofluorocarbon solvents (CFC-13), which deplete the ozone layer [144]. For instance, the boiling point of HFC-134a is −25°C and this compound is described as a nonflammable and nontoxic substance.

The solvent also has a moderated vapor pressure of 5.6 bar (= 4.93 atm) at ambient temperature; thus, in general terms, it could be considered among most standards a poor solvent, with limited solvation activity for several polar and nonpolar raw materials [144, 145].

Because of these physical chemistry limitations, the solvent does not mix with mineral oils or triglycerides and does not dissolve plant wastes.

Nonetheless, the process has its advantages:

- It can be customized by using modified solvents based on HFC-134a.
- By using HFC-134a-modified solvents, it is possible to extract a broader scope of components.
- The process can be highly selective for the extraction of specific classes of phytoconstituent(s) in a plant sample.

Also, natural products extracted through this process have shown to have extremely low residual solvent when compared with other conventional methods; therefore, the residuals detected after the samples are dried invariably presented less than 20 parts per billion (ppb) and are frequently below common standards of analytical levels of detection [146].

In terms of experimental advantages, these selected fluorinated solvents are chemically neutral regarding their pH activity because they are neither acidic nor alkaline and, therefore, have only minimal potential collateral reaction effects on the extracted botanical materials. However, the advantages credited fluorinated solvents has for chemical extraction, presents equipment limitations due to the process needs to be carried out on a totally sealed chamber to avoid solvent's evaporation and although extraction can be carried out at room temperature, to afford completion, it needs to be continually recycled with product fully recovered at the end of each cycle.

Because of these unique properties, it is thought that the only utility needed to operate these extraction systems is "power" and, even then, the consumption of energy is also hypothesized to be low. Therefore, there is no scope for the solvent(s) to escape when an extraction is running at room temperature, and even if some volumes of solvent do escape, they contain not a single chlorine molecule and, therefore, do not pose threat to the ozone layer.

At the completion of an extraction, the residual waste biomass from an extracted plant is a material commonly recovered as dry and "eco-friendly" to be handled in further steps [145].

Proposed advantages of the process

- Unlike other processes that employ high temperatures, the phytonics process is cool and gentle and its products are never damaged upon exposure to temperatures more than the ambient temperature.
- No vacuum stripping is needed, which, in other processes, leads to the loss of precious volatiles.
- The process is carried out entirely at neutral pH, and in the absence of oxygen, the products never suffer acid hydrolysis damage or oxidation.
- The technique is highly selective, offering a choice of operating conditions and hence a choice of end products.
- It is less threatening to the environment because of the absence of chlorine.
- It requires a minimum electrical energy consumption.
- It releases no harmful emissions into the atmosphere, and the resultant waste products are innocuous and pose no effluent disposal problems.
- The solvents used in the technique are not flammable, toxic, or ozone depleting like chlorofluorocarbons.

- The solvents are completely recycled within the system.
- Low traces of solvent are detected in the final products.

Concerns and disadvantages

- By most liquid standards, HFC-134a is a poor solvent.
- The solvent does not mix with mineral oils or triglycerides.
- The solvent does not dissolve plant wastes.
- The technique only operates under sealed conditions of the extractor.

Although this technology is argued to be in many ways most affordable and reliable than other established ones, not so much information is available until now in the extraction literature that allows to stablish robust evidence on the benefits related with this extraction technique [146–151].

Chapter 5

Critical Extraction Methods

5.1 Introduction

In parallel with society's needs, new extraction technologies and methods have been designed for the rapid and reliable extraction/separation of valuable bioproducts by utilizing innovative processes such as ultrasonic, microwave, membrane separation, molecular distillation, macroporous resin adsorption, phytonics, hydrothermal, and supercritical fluid extraction, with a great impact on the extraction of effective components that are found in medicinal plants, by promoting the development of plant extracts in the fragrance, pharmaceutical, nutraceutical, and food industry.

From these emergent technologies, supercritical fluid extraction, mostly referred to as SFE, is the most compelling one as it can be employed for heat-sensitive molecules. It has its own unique advantages, including low to nontoxicity, no-solvent residue, high extraction rate, and free of environmental pollutants.

Nonetheless, like every coin, it has two sides. One issue is that SFE has less sample capacity, mainly for scaling up for the extraction of high-value-added bioproducts. Therefore, its industrial application has been limited because the technology is only efficient when operated at high pressures by doing its scale-up costliest when compared to other non-critical feature designs such as maceration, percolation, Soxhlet, and semi-critical within others.

Semi-Critical Assisted Extraction: Applications and Commercialization in Biotechnology, Food, and Pharmacy
Tulio Chavez-Gil

ISBN 978-981-4968-36-2 (Hardcover), 978-1-003-29124-4 (eBook)
www.jennystanford.com

In parallel with SFE development, another extraction technique, "subcritical water extraction" or "hydrothermal extraction," also emerged. The subcritical process, however, has the same physical principles, advantages, and applications like those presented by the SFE for the extraction of a range of medicinal plants, traditionally used in the United States, Asia, and Europe to extract pharmaceutical compounds (e.g., semi-volatile to nonvolatile oils, dyes, antioxidants, analgesic-alkaloids, nutraceuticals, etc.), because the equipment and experimental parameters for both methods are a bit similar and will be discussed in this chapter.

5.2 Pure Substances: The Critical Point's Relevance

5.2.1 Pure Substance

A substance that has a fixed chemical composition throughout is called a pure substance. Typical examples such as argon (Ar), carbon dioxide (CO_2), nitrogen (N_2), and water (H_2O) can all be defined as pure substances. Nonetheless, a pure substance does not have to be a single chemical element or compound, altogether. Therefore, a mixture of either pure chemical elements or compounds also qualifies to be classified as a pure substance if the mixture is shown to be homogeneous. Air, for example, is a mixture of several gases, but it is often considered a pure substance because it has a uniform chemical composition (see Fig. 5.1).

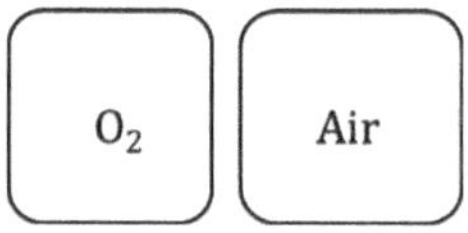

Figure 5.1 An oxygen (molecule) and gaseous air (a compound) both are pure substances.

On the other hand, a mixture of oil and water does not classify to be considered a pure substance because oil is not soluble in water. It is collected on the top of water, forming two chemically dissimilar phases due to their polarity differences, as discussed in Chapter 2. Also, a mixture of two or more phases of a pure substance is still a pure substance if the chemical composition of all the phases remains the same (as seen in Fig. 5.2).

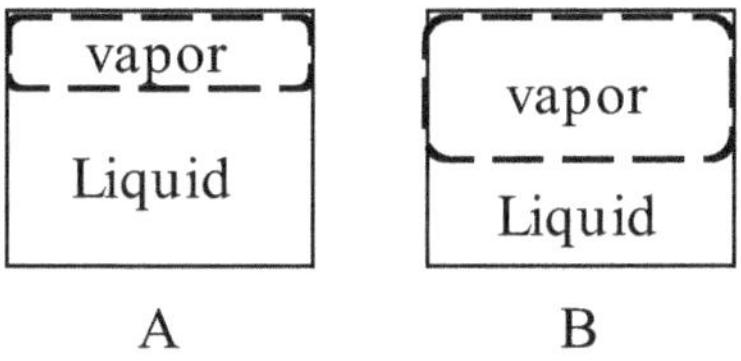

Figure 5.2 (A) A mixture of liquid and gaseous *water* is a pure substance, but a mixture of liquid and gaseous *air* is not (B).

A mixture of ice and liquid water, for example, is a pure substance because both phases have the same chemical composition. However, a mixture of liquid and gaseous air collected in the same container cannot be classified as a pure substance since the composition of liquid air is different from that of gaseous air, and the mixture is no longer chemically homogeneous if some temperature is applied, because different components found in air condense at different temperatures and at specified pressures.

5.2.2 Thermophysical Conductivity of Pure Substances

When a pure liquid is poured inside a glass tube and sealed, the equilibrium achieved with its vapor at a constant temperature T_1 is a bit perturbed if the liquid is subjected to a gravitational field, with a little change observed at the top of the liquid due to the formation of a concave interface, the so-called "meniscus." This situation corresponds to the isotherm T_1 in Fig. 5.3, where the point 2 represents the vapor-pressure equilibrium of liquid and corresponds to the liquid's molar volume, v, whereas the point 1 corresponds to the larger molar volume of the vapor phase at the same pressure and temperature.

If no changes are introduced on P and T, both the points 1 and 2 lie on the vapor-pressure curve, which separates the liquid and vapor phases, because the expansion coefficients have different values for a pure substance when it is found in two different physical phases; therefore, the same substance is found predominantly in one state instead of the other as the functions of both P and T change. If, for example, temperature is raised in the sealed tube containing a pure liquid to T_2 (Fig. 5.3), its vapor pressure is increased along the projection v_C (in Fig. 5.4). At the same time, the molar volume of the liquid phase is increased, while that of the vapor is decreased, as shown in the diagram of that figure. If this process is continued, both liquid and vapor phases eventually become indistinguishable at the same temperature (the critical temperature), which is the situation indicated by the *critical points* in Figs. 5.3 and 5.4, respectively. At this critical temperature T_c, the meniscus would disappear and the molar volumes of the two phases are equal to v_1, v_2 as depicted in Fig. 5.4. In addition, the critical temperature of a pure liquid may also be defined as the highest temperature at which the liquid phase of the pure substance can exists. In this regard, when thermophysical dimensions (P, T) determine the behavior that a pure liquid substance exerts during a chemical process, the direct relationship between the pressure P of the substance to its molar volume v, at constant temperature T, will follow the projections depicted in Fig. 5.3, while the projections of the same behavior with P and T as experimental coordinates and volume v as the parameter derivative from the changes introduced by both P and T are represented in Fig. 5.4, respectively.

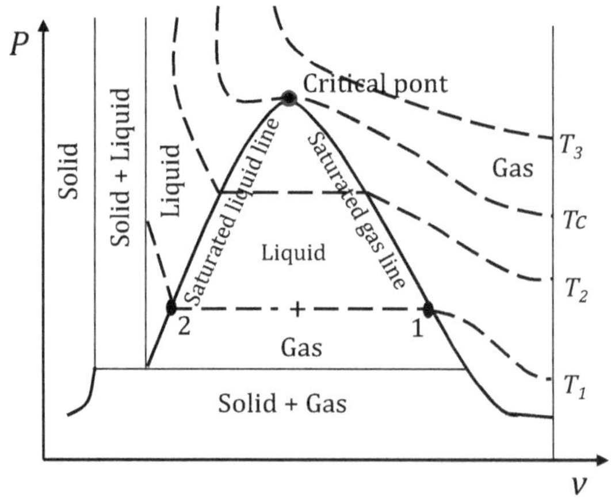

Figure 5.3 *P–V* diagram of a pure substance as a projection of its *P–T* behavior.

As can be seen from the *P–V* diagram in Fig. 5.3, along the critical isotherm T_c, the relationship between the fluid pressure and its molar volume shows a point of inflection at the molar volume v, the critical volume v_c, and the corresponding critical pressure p_c, so that the conditions in Eq. 5.1 define the critical point as:

$$\frac{(\partial p)}{(\partial v)_{Tc=0}};\quad \frac{(\partial^2 p)}{(\partial v^2)_{Tc}}=0;\quad \frac{(\partial^3 p)}{(\partial v^3)_{Tc}}<0 \tag{5.1}$$

It is the unique thermodynamic state for which at temperature T_c, the molar volume of the pure substance is v_c, and its pressure is p_c.

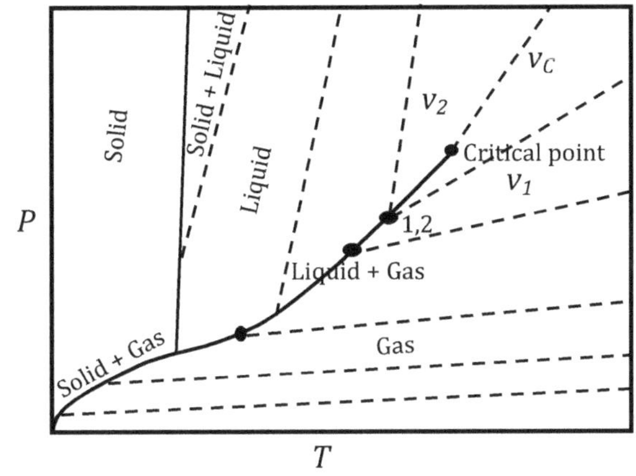

Figure 5.4 Diagram of the *P–T* projection of the *P–V* behavior of a typical pure substance.

Nonetheless, it is necessary only to describe two of these critical state parameters since the third is then automatically determined from the other two. By taking into consideration this physical reasoning, the two critical state parameters T_c and p_c are, therefore, assumed as characteristics of any pure substance, and then those must be determined experimentally. Though some other physical properties of pure fluids exhibit special behavior near its critical point, as for example when the heat capacity of a fluid becomes infinitely large at the critical point as does thermal conductivity, whereas its thermal diffusivity becomes zero [156, 157].

This behavior is very important when "efficiency" is the key criterium to consider if the thermal conductivity of a pure substance is used to evaluate technologies such as critical, subcritical, and supercritical extraction by assuming their performance rests on the thermophysical properties that characterize carbon dioxide near its critical point. Though critical effects are the result of large-scale fluctuations in the density of the fluid, which extends over a wide range of temperatures and volume around the critical point. For example, the effect upon the isobaric heat capacity and thermal conductivity of CO_2 can be more than 10%, even 50 K above the critical temperature at the critical density as observed in Fig. 5.5 [157]. In general, a pure solvent achieves firstly a biphasic equilibrium when it is placed in a sealed container and free of external activity or if a pure gas is dissolved in it in a similar container as observed in the diagrams of Fig. 5.2A,B. Nonetheless, if the solvent is placed in an open or large space, a triphasic equilibrium is observed at pressure and temperature up its freezing or melting but below its boiling points with each phase characterized to has different specific volume as exemplified in the projections of Fig. 5.4.

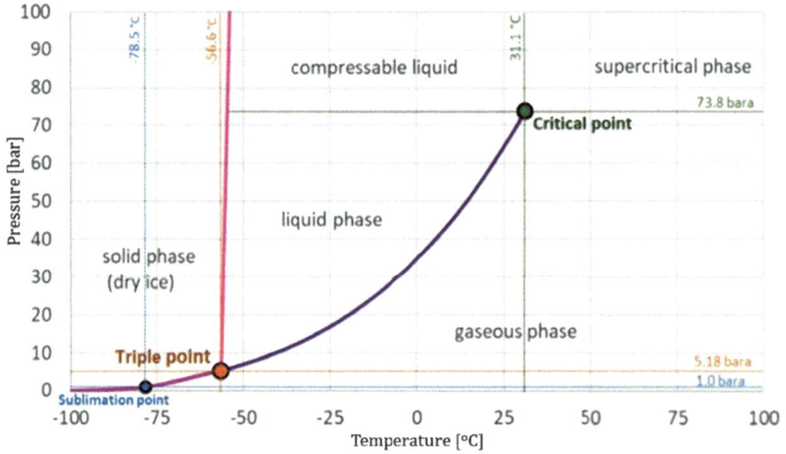

Figure 5.5 Diagram of thermophysical properties of carbon dioxide, CO_2. Reprinted from Ref. [157].

The triple line appears as a point on the *P–T* diagrams (Figs. 5.3 and 5.4); therefore, it is often called the **triple point**. In this regard, the triple-point temperature and pressure of various substances that are used as common solvents in chemical extraction are given in Table 5.1. For water, in particular, the triple points related with temperature and pressure are 0.01°C and 0.6117 kPa, respectively.

Table 5.1 Temperature and pressure correspondent to the triple point of pure solvents used in chemical extraction[a]

Substance	Formula	T_{tp} (K)	P_{tp} (kPa)
Acetic acid	$C_2H_5O_2$	594.45	58785.7*
Acetylene	C_2H_2	192.4	120
Ammonia	NH_3	195.40	6.076
Argon	Ar	83.81	68.9
n-Butane	C_4H_{10}	134.6	7×10^{-4}
Carbon (graphite)	C	3900	10,100
Carbon dioxide	CO_2	216.55	517
Carbon monoxide	CO	68.10	15.37
Chloroform	$CHCl_3$	175.43	0.870†
Deuterium	D_2	18.63	17.1
Ethanol	C_2H_6O	150	4.3×10^{-4}‡
Ethane	C_2H_6	89.89	8×10^{-4}
Ethylene	C_2H_4	104.0	0.12
Helium 4 (λ point)	He	2.19	5.1
Hydrogen	H_2	13.84	7.04
Hydrogen chloride	HCl	158.96	13.9
Iodine	I_2	386.65	12.07†
Isobutane	C_4H_{10}	113.55	1.948×10^{-5}†
Mercury	Hg	234.2	1.65×10^{-7}
Methane	CH_4	90.68	11.7
Neon	Ne	24.57	43.2
Nitric oxide	NO	109.50	21.92
Nitrogen	N_2	63.18	12.6
Nitrous oxide	N_2O	182.34	87.85
Oxygen	O_2	54.36	0.152
Palladium	Pd	1825	3.5×10^{-3}
Platinum	Pt	2045	2.0×10^{-4}
Sulfur dioxide	SO_2	197.69	1.67
Titanium	Ti	1941	5.3×10^{-3}
Uranium hexafluoride	UF_6	337.17	151.7
Water	H_2O	273.16	0.61
Xenon	Xe	161.3	81.5
Zinc	Zn	692.65	0.065

Data adapted from [a]National Bureau of Standards (US) Circ., 500 (1952), and Refs: *[158]; ‡[159]; †[160]; [161]

Thus, all three phases (or states) applied to water coexist in thermophysical equilibrium only if the temperature and pressure have precisely these values. As a broad assumption applied for liquid solvents, a pure substance cannot exist in the liquid phase at a pressure below its triple-point pressure. This same reasoning can be given for the temperature dependence of substances that contract at their freezing points. Therefore, a pure liquid substance subjected to high pressures can exist in the liquid phase even at temperatures below its temperature's triple point.

In the case of pure water, however, the substance cannot exist in the liquid phase in equilibrium with the other two phases at atmospheric pressure (101.325 kPa) and temperatures below 0°C, but it can exist as a liquid at 20°C under 200 MPa of pressure.

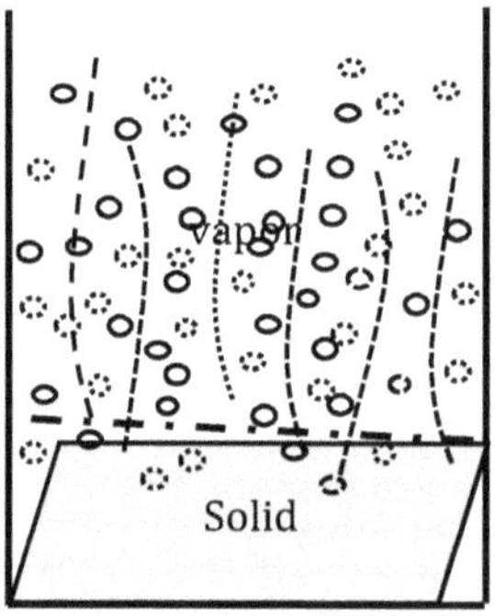

Figure 5.6 Scheme showing how at low pressures (below the triple-point values), solids evaporate without melting first (sublimation).

Furthermore, pure water in its solid phase (ice) exists at seven different solid phases at pressures above 100 MPa and 0°C [161]. There are two ways in which a substance can transition from the solid to the vapor phase: either the solid melts first into a liquid and subsequently evaporates, or the solid evaporates directly without melting first. The latter change, however, only occurs at pressures below the triple-point values since the pure substance cannot exist in the liquid phase at the pressures it is found as solid (Fig. 5.6). Thus, for those substances characterized for an easy physical transition from their solid phase directly into the vapor phase at standard conditions (0–20°C, 1 atm), the change is called **sublimation**. A substance typically associated with sublimation, which has its triple-point pressure above the atmospheric pressure, is solid carbon dioxide (CO_2) (dry ice), and sublimation is the only thermophysical way this pure substance can change from the solid to the vapor phase at standard conditions (273 K, 1 atm).

5.2.3 Thermophysical Changes *The P–v–T Diagram*

The physical state changes experienced by a pure and compressible substance such as a liquid hydrocarbon or a gas molecule can be fixed by any of the other two independent and intensive physical chemistry dimensional properties. Once the two independent physical dimensions are thermodynamically fixed, all other physical or chemistry properties become dependent properties from the first two [161].

It will be useful at this point if we remember that any equation with two independent variables in the form $z = z(x, y)$ represents a surface in the space, so we can draw for the thermodynamic properties the *P–v–T* behavior of a pure substance, as a surface in space, as shown in Figs. 5.7 and 5.8, respectively. In the diagrams, *T* and *v* dimensions may be analyzed as the independent variables (the horizontal lines), while *P* is viewed as the dependent variable (the ordinate). All the points on the surface represent the phase equilibrium state. Besides, all states occurring along the path on a quasi-equilibrium process lie on the *P–v–T* surface since such a process must pass through equilibrium states.

In the single-phase regions (gas, liquid, solid) in Fig. 5.8, it appears as curved surfaces on the *P–v*–T diagram and the two-phase regions (liquid–vapor, and solid–vapor) appear as surfaces perpendicular to the *P–T* plane. This behavior is expected as the projections of the two-phase regions on the *P–T* plane are perpendicular lines. Nonetheless, all the two-dimensional diagrams that have been discussed so far are merely projections of the three-dimensional surface onto the appropriate planes.

For example, the *P–v* diagram is just a projection of the *P–v–T* surface on the *P–v* plane, and a *T–v* diagram is nothing more than a naked eye view of this proposed surface. Therefore, the *P–v–T* surfaces present a great deal of information at once, but for purposes of thermodynamic analysis, it is more convenient to work with two-dimensional diagrams, such as the *P–v* and *T–v* diagrams only.

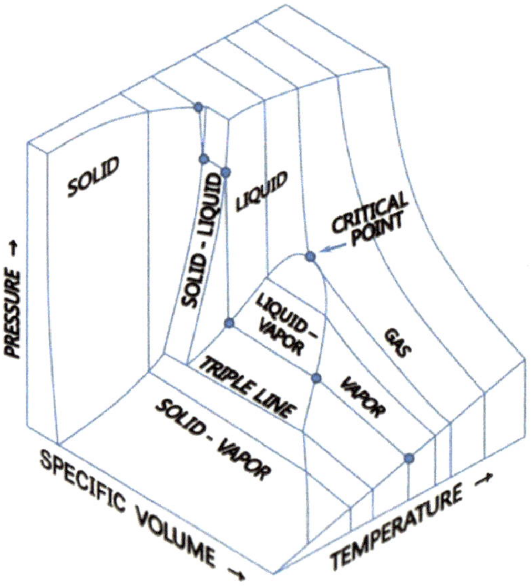

Figure 5.7 A *P–v–T* surface diagram for a pure substance that *contracts* on freezing. Modification adapted from Ref. [161].

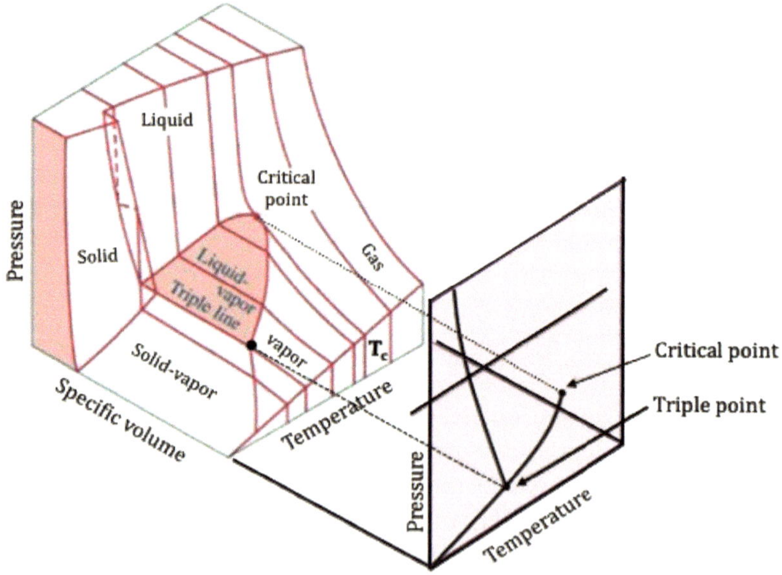

Figure 5.8 A *P–v–T* surface diagram of a pure substance that expands at freezing conditions like water. Modification adapted from Ref. [161].

5.2.4 Thermomechanical Extraction: Critical Strategies

5.2.4.1 Instant controlled pressure drop technique (DIC)

Very recently, the instant controlled pressure drop method, called DIC (*Détente Instantanée Contrôlée*) in French, has been introduced by Allaf et al. Since its research conceptualization, the technique has been widely used for extracting various kinds of natural raw materials possessing various chemical constituents of medicinal and pharmaceutical interest [162].

As a matter of fact, the DIC method has been constantly improved and mended in the domain of upgraded extraction methods for multiple industrial applications. The method is characterized to operate at high temperatures and high pressures in a relatively short time by exhibiting mainly thermomechanical effects on its target material. The thermomechanical effects are created after the raw materials are subjected for a short period of time to a saturated fluid steam, followed by an abrupt pressure drop by leading the system to an either moderated or high vacuum. A typical experiment is carried out on a previously humidified sample that is placed in the extraction vessel, as seen in Fig. 5.9, and is subjected to an initial vacuum step. At vacuum, the system is heated under high temperature (up to 180°C) during a short time (5–60 s) and the system's pressure is saturated to 1 MPa. At this point, the pressure is abruptly dropped to achieve the vacuum (3–5 kPa) that occurs at almost 20 to 200 ms. It is thought that applying an abrupt pressure drop ($\Delta P/\Delta t >$ 25.106 Pa/s) may cause critical mechanical stress on the cell wall by inducing a process of auto-vaporization of fluids (like water), which is followed by an instantaneous cooling of the sample [162, 163].

This instantaneous pressure change is responsible for a quick swelling of the matrix, which occurs simultaneously and may lead to the rupture of the matrices with the subsequent secretion of metabolites through the breaking walls [164].

The operation sequence that allows this thermodynamic phenomenon may aid in the classification of the DIC method as a "green extraction" technique (see Fig. 5.9) for the utilization of steam water as a powerful solvent, which would result in a reduced consumption of organic solvents when compared to employing traditional extraction methods [162, 165].

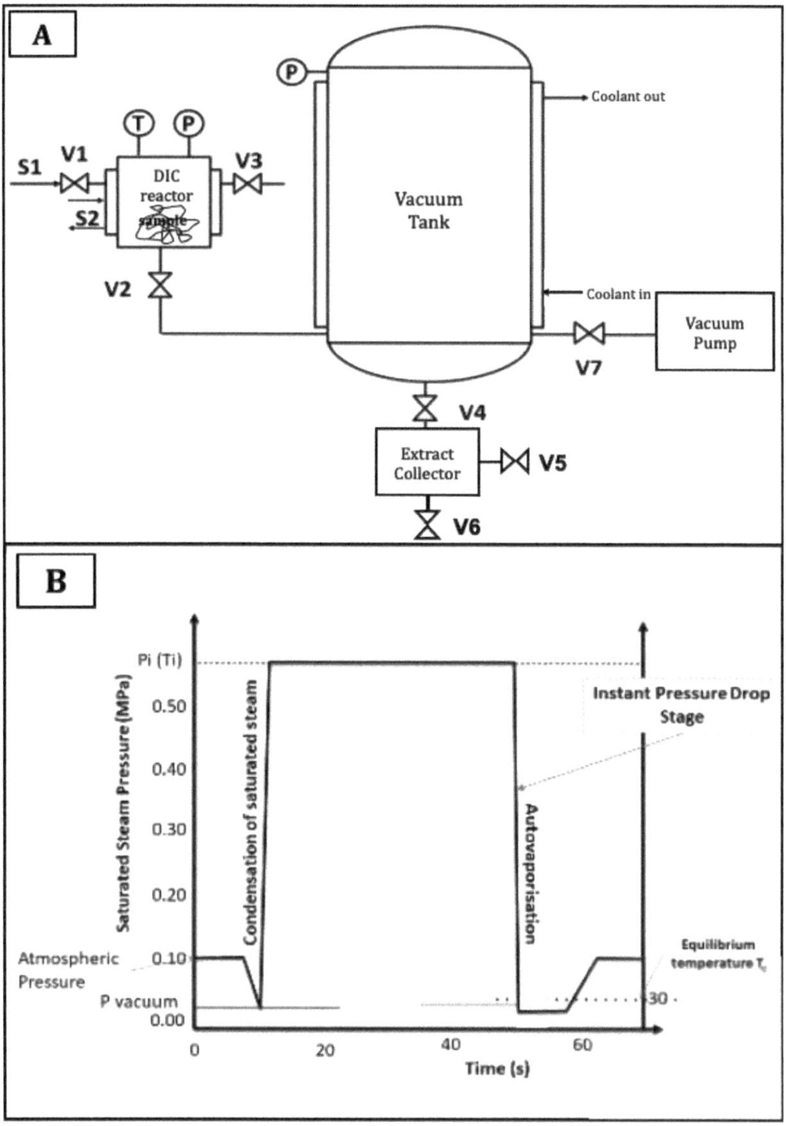

Figure 5.9 Simplified scheme of DIC extractor (A), and setup parameters diagram for DIC extraction (B). Adapted and modified from Ref. [162].

The DIC green extraction technology occurs in an engineering design of four main components (as seen in Fig. 5.9):

1. An extraction vessel where the sample to be treated is placed inside a vessel possessing a heating jacket that acts as an autoclave.
2. A valve that connects the steam generator and the vacuum pump with the extractor vessel and is operated to take control of the instant negative pressure and to ensure the control of positive pressure induced by steam water as the extractant inside the reactor.
3. A vacuum system located below the extraction vessel comprising a vacuum pump and a tank with a cooling jacket.
4. A collection tank to recover the extracted condensates.

5.2.4.2 Using DIC pretreatment for sustainable extraction

The DIC method for the pretreatment of solvent extraction has been considered an efficient technology to enhance the chemical activity of a solvent before the extraction per se. The enhancement activity is credited to have action by improving solvation, which leads to the expansion of cell matrices by softening the texture of the plant tissues for high solvent activity and creating better porosity on the surface of cell walls. In a recent publication, Mkaouar et al. reported on the efficacy of the DIC method after the post-treatment of olive leaves (*Olea europaea*), which showed a steep enhancement when the leaves were subjected to ethanol solvent extraction of their phenolic compounds [165]. This research group reported that the extraction yield augmented by 312% and the extract was richer in bioactive compounds. Similar reports on the DIC pretreatment method have shown positive results and significant cell disruption effect on the extraction of different phenolic compounds from grape stalk powder conducted with solvents characterized by presenting different polarity [165, 166]. Another study proved that DIC promoted the enhancement of lipid extraction from rapeseed and jatropha seeds without significant modification of fatty acids composition when compared with the analytical composition through conventional Soxhlet extraction [167, 168]. A study conducted by Allaf et al. showed that lipid extraction could be enhanced by DIC pretreatment, and their results agree with the statistical calculation for the effective diffusivity of solvent inside the cell matrices (see Table 5.2) [162, 168]. More recently, Eikani et al. reported that the DIC method

Table 5.2 Oil yields from camelina seeds and pressing-meals

Run no	**$DIC_{1;4;10;13}$**	**DIC_2**	**DIC_3**	**DIC_5**	**DIC_6**	**DIC_8**	**DIC_9**	**DIC_{11}**	**DIC_{12}**	**RM**
Saturated Steam Pressure (Mpa)	0.45	0.70	0.45	0.63	0.63	0.27	0.27	0.20	0.45	—
Processing Time (s)	70	70	120	105	35	35	105	70	20	—
$Y_{ASE;seeds}$	588.0±6.0	603.0	610.4	615.9	570.0	569.4	571.4	573.9	575.3	555.50± 4.00
$Y_{2h\text{-}DM;seeds}$	451.0±1.0	455.8	442.6	462.8	453.6	439.9	439.5	446.5	441.0	284.80±0.05
$Y_{pressing}$	475.0±1.0	484.7	485.5	490.9	463.3	449.8	481.7	472.1	451.8	444.70±4.00
$Y_{ASE;meals}$	131.8±0.2	133.3	132.4	133.4	133.0	131.0	131.2	130.7	131.7	110.00±0.25
$Y_{2h\text{-}DM;meals}$	113.0±4.0	114.1	113.8	115.8	114.5	111.4	112.5	111.9	113.6	51.60±0.27

$Y_{ASE;seeds}$: Accelerated solvent extraction oil yield from camelina seeds assisted by *DIC*
$Y_{2h\text{-}DM;seeds}$: Dynamic Maceration 2h oil yield from camelina seeds assisted by *DIC*
$Y_{pressing}$: Pressing extraction 1 oil yield from camelina seeds assisted by *DIC*
$Y_{ASE;meals}$: Accelerated Solvent Extraction oil yield from camelina meals assisted by *DIC*
$Y_{2h\text{-}DM;meals}$: Dynamic Maceration 2h oil yield from camelina meals assisted by *DIC*
RM: Raw material
Reprinted from Ref. [126].

is an efficient pretreatment technology for the intensification of bio-oil extraction from safflower and castor seeds [169, 170]. The increased interest in the DIC pretreatment technology applied for green extraction has been summarized in recent publications on the work performed in the last 10 years [171–183]. Thus, this extraction technology has received great attention and a consistent improvement is observed in recent times as aforementioned in this section, specially, on technological and scientific bottlenecks by overcome, often up to 50% on investments for scaling up and more than 70% of the efforts to reduce energy, solvent, and increase production [176]. In this direction, new methods considered more environment friendly, safe, less time consuming, highly efficient, and economically profitable among green solvents are currently a matter of intense discussion in the scientific community [176–183].

5.2.4.3 DIC for green extraction: solvent advantages

Nowadays, the field of chemical extraction has entered the so-called "green zone" revolution after Chemat et al., in 2012 [164], introduced the concept of "green extraction of natural products" based on the "green chemistry" advantages, further referring to modern sustainable processes which involve "green solvents" and are hypothesized works on the bases of both new solvent discoveries accompanied by the design of less-to-short extraction steps as synergic mechanisms that leads to reduce energy consumption, use solvents other than halogenated or petroleum derivatives in an accelerated needs to handle a wide scope of renewable natural sources for the extraction of more safe and high-quality extracted bioproducts.

Nonetheless, the implementation of new ideas that address the challenges encountered with old methods requires that a proposed innovation breaks away from the past, instead of simply improving yields but continuing the usage of traditional solvents and the same extraction devices.

In this regard, the DIC extraction method can be operated under extreme or non-conventional conditions. It is being constantly improved, and currently a dynamic developing extraction area is under exploration, to be used in applied research and industrial experiments.

By using green extraction techniques, researchers and technicians have completed processes that can now be ensured in a few minutes instead of hours, accompanied by high reproducibility, high investment benefits, low consumption of solvent, simplified operation, possessing analytical advantages such as higher purity of the final product(s), low investment for the elimination of post-treatment of wastewater, and consumption of only a fraction of energy normally needed by using a conventional extraction method [171]. Researchers involved in the "green extraction" technology invite readers like chemists, biochemists, chemical engineers, physicians, food, and biotechnologists working even in academia or industry to find major solutions to problems related to extraction's design by demonstrating how sustainable this processing technique is for the laboratory, classroom, and industrial scale-up regarding optimal transformation of raw food materials, water, and energy in the ways to (1) improve and optimize current existing processes, (2) using non-dedicated equipment, and (3) innovation to improve processes and procedures [172–176]. The green extraction technology uses supercritical solvents as a replacement of conventional organic solvents in laboratory processes that include different industrial products such as food, pharmaceuticals, agriculture, and cosmetics. Many compounds have been considered supercritical fluids, for example hydrocarbons (pentane, butane, hexane), aromatics (benzene, toluene), alcohols (methanol, isopropanol, n-butyl alcohol), and some gases (carbon dioxide, ethylene, propane) [177–183]. However, among the above-mentioned supercritical solvents described, CO_2 is unquestionably the most often employed solvent due to its numerous benefits.

A fluid is regarded as supercritical when its pressure and temperature are beyond its critical points, that is, the solvent's critical pressure (T_c) and critical temperature (P_c). In this physical point, the solvent is an undistinguishable fluid that forms a biphasic state, where it is even found as a gas and liquid in an absolute equilibrium, which gives it unique physical chemistry properties in an advantageous way with their chemical activities very close to that of a gas in some instances, but also close to liquid in others [185].

Table 5.3 Chemical solvents most used in critical and supercritical fluid extraction (SFE) and their critical characteristics

Solvent	Mol. Weight [g/mol]	Crit. Temp [K]	Crit. Press [MPa]	Crit. Density [g/cm^3]
CO_2	44.01	304.1	7.38	0.469
Water	18.02	647.3	22.12	0.348
Methane	16.04	190.4	4.60	0.162
Ethane	30.07	305.3	4.87	0.203
Propane	44.09	369.8	4.25	0.217
Ethylene	28.05	282.4	5.04	0.215
Propylene	42.08	364.9	4.60	0.232
Methanol	32.04	512.6	8.09	0.272
Ethanol	46.07	513.9	6.14	0.276
Acetone	58.08	508.1	4.70	0.278

Modified from Ref. [184].

In the case of CO_2, for example, a simple phase diagram can demonstrate the concept of the supercritical state. Thus, the critical point of CO_2 usually exists in the three classical physical states (solid, liquid, and gas) due to variations in pressure and temperature [157], but sometimes a fourth CO_2 phase is found, "the supercritical fluid," as the result of a dynamic equilibrium that appears at the point where there are no differences between its liquid and gas phases [186]. The phase diagram in Fig. 5.5 shows the different points and physical states of CO_2 in terms of combinations of the dependent variables occurring at an either high or low temperature and pressure as discussed above. In the diagram, the point that represents all the three phases of CO_2 together at equilibrium (−56.6°C and 51.1 bar) corresponds to its triple point.

The critical point, on the other hand, appears at 31.1°C and 78.3 bar, which are the T_c and P_c of CO_2, as discussed in Section 5.2.2. The region above the critical point is described as the supercritical fluid region, which appears when there are no physical differences between a gas and a liquid, and over that critical point, no matter how much pressure or temperature is increased, the substance cannot be changed from liquid to gas or from gas to liquid in an isoequilibrium and isokinetic relationship [187]. In general, supercritical fluids are characterized by display physicochemical properties between those behaving as a liquid and gas, typically characterized by low viscosity, high density, and high diffusivity, which can be managed by simple changes in their pressure and/or temperature (see Table 5.4) [188]. Those are the thermophysical properties mainly dominant during the use of a supercritical solvent that makes them excellent fluids for different applications due to the disruptive power of their cell matrices, which are based on their high mass transfer rate and high activity on natural fluids that possess high density in the extractable components [189].

Table 5.4 Comparison of physicochemical properties of liquids, gases, and supercritical fluids

State	Density [kg/m^3]	Viscosity [μPa]	Diffusivity [mm^2/s]
Liquids (P=1 atm, T=15–30°C)	1000	500–1000	0.001
Gases (P=1 atm, T=21°C)	1	10	1–10
Supercritical fluids ($P=P_c$, $T=T_c$)	100–1000	50–100	0.01–0.1

Reprinted from Ref. [184].

5.2.4.4 Supercritical fluids: physical characteristics

The density of a supercritical fluid is thought to be between that of a gas and a liquid, but the consensus is that it is closer to that of a liquid. Therefore, it mostly relies on pressure and temperature conditions because if pressure rises at a constant temperature, it increases the density of a supercritical fluid, and then its solvating power by default. On the other hand, when pressure remains constant and temperature rises, the process reduces the density of the supercritical fluid together with the solvent strength [186]. Therefore, these are the main thermodynamic characteristics and fundamental physical sources responsible for the excellent solvation properties of a supercritical fluid when it is acting on a solute molecule, which presents strong chemical characteristics.

The viscosity of a supercritical fluid is one of the fundamental thermophysical properties of a supercritical solvent that is generally known to be low and nearly equal to that of the gas state, but less than that of the fluid at its liquid phase [189]. When a solvent characterized to have low viscosity is employed in an extraction process, it may enhance the solvation power, whether it is a supercritical solvent due to the lower resistance encountered on the cell walls than that observed by a conventional liquid acting on a natural solute.

At this point, we can state that temperature has a higher impact on the viscosity of supercritical solvents than on the viscosity of common liquids [187].

The diffusion activity, on the other hand, is typically higher for supercritical fluids than that observed in a liquid but is lower when compared with a gas [190]; consequently, a solute can show better diffusivity in a supercritical solvent than in a liquid, whether pressure is constant, and temperature raised. Therefore, analyzed together, temperature and pressure conditions lead to different effects when applied on gases, but the effect of these physical conditions is straightforward if supercritical solvents are involved in the same experiment: for example, if pressure increases, the diffusivity of a supercritical solvent decreases and when the temperature increases, the diffusivity increases. In this case, the diffusivity property enhances the extraction capacity of supercritical solvents when compared with a liquid solvent; therefore, they are more suitable solvents for analytical purposes [187, 188].

5.2.4.4.1 *Temperature and pressure effects: advantages/disadvantages*

Temperature and pressure are the primary parameters that influence the extraction efficiency because of their direct effect on the solubility of a substance by the activity of a supercritical solvent.

Experimentally, an increase in pressure at a specific temperature in the process increases the density of the supercritical solvent and then the solubility of targeted compounds [191]. That means the higher the pressure, the solvent volume decreases, and then a less volume of solvent is needed for a particular extraction [192]. Conversely, by elevating pressure to a given point, this action can reduce the solvent activity because its density is a bit increased and a decrease in its diffusivity is observed, which reduces the solute's solubility [193]. Furthermore, the application of high pressure is not recommended for the extraction of all substances and targeted compounds, because this procedure can lead to a total compact of the raw material, which can adversely affect the extraction yield [194].

Similarly, the extraction temperature occurring at constant pressure has two opposing effects on supercritical solvents. First, at increased temperatures, a decrease in the solvent density is introduced, thus reducing its solvating power, but this action has effects on improving the vapor pressure of some desired compounds, and consequently increasing the analyte solubility and the extraction yield. But as mentioned earlier, an increase in temperature decreases the solvent density and, consequently, may reduce its solvating power and the solute solubility by negatively affecting the extraction yield [195]. An analysis of these two opposite temperature effects (solvent's solvation/solute's solubility) may result in a crossover isotherm in a phenomenon known as "retrogradation" in which, a liquid that is turned into a gel by decreasing its temperature makes that a linear molecule solubility decreases as consequence of the density changes on that solvent [196]. In this case and according to the literature, the temperature of a supercritical fluid generated during a supercritical extraction applying for thermolabile compounds must be fixed between 35 and 60°C to avoid degradation [197] processes, and the pressure should be fixed to 400 bar to avoid an increase in the solvent's density. Therefore, these thermophysical conditions should be carefully selected based on the purpose of the process and the nature of the targeted chemicals [198]. Because pressure and temperature are experimental properties connected spatially and considered responsible for the extraction's efficiency, it makes supercritical fluid extraction (SFE) stand out as an excellent thermophysical technology. By employing SFE, for example, the induced low viscosity and high diffusivity of solutes in the solvent allow better mass transport properties than those achieved through cold liquid activity observed through a conventional Soxhlet process.

Thus, supercritical solvents diffuse effortlessly in a solid sample by increasing extraction efficiency and product yield in a short time. In addition, improved density combined with solvent power activity can be tailored automatically because these induced conditions can give high solubility among high selectivity to the SCF technology; therefore, a large number of new compounds can be extracted to be employed in the development of new products like nutraceuticals and fine chemicals [199]. Selected studies performed with the aim of understanding the influence of either temperature or pressure or both on supercritical solvents during the extraction of bioactive substances offered different results due to the choice of plant materials as well as the targeted compounds selected [199–203]. Different groups performing independent experiments showed that high pressures were responsible for the enhancement in the total extraction yield of phenolic compounds [200, 202]. For example, an experiment reported that phenolic groups were determinate by submitting the supercritical solvent to low pressures with no significant differences in the yield product when compared to those where substantial increases were done on both temperature and pressure [201]. Nonetheless, a number of published results showed that a substantial augmentation in the recovery of phenolic compounds is observed on reducing the experiment's temperature and consequently an increase in the addition of a co-solvent as well as pressure [200].

5.2.4.5 Natural bioactive compounds extracted by SF

An increased interest in supercritical extraction/separation technologies for renewable sources application had attracted considerable attention because the confirmed effects that supercritical solvents has on targeted chemical groups and makes that some of these upgraded technologies shown to be more efficient and reliable in some aspects than the empiric methods (Soxhlet, maceration, percolation) for industrial production of supplemental food, fertilizers, animal feed, biofuels, medicines, cosmetics, nutraceuticals, and fine chemical feedstocks, discussed in former chapters. These technologies are granted to be less hazardous due to the solvents used in them, fulfill safety regulations because the operators' scholar preparation, show some opportunity for scaleup, and all of them are classified as "green technologies" because are environmentally friendly.

From the science and biotechnology point of view, medicinal, vegetable, and food plants are the broad resources of chemical components that had been historically used as natural remedies for some chronic and opportunistic diseases. Now the main components of even wild and farm-grown plants are used as food additives (e.g., natural flavor, aroma, and color), in the production of functional foods, in agriculture as natural pesticides, and for the development of new drugs [204].

The bioactive compounds of plants are classified as natural molecules with non-nutritive components of plant food biosynthesized as secondary metabolites [205]. Medically, they can only provide essential benefits or undesirable effects for human or animal cells and are found in small to moderate concentrations in plant cell matrices; therefore, they are biologically valuable and technologically significant for extraction, isolation, characterization, and economic exploitation [206].

Furthermore, plant bio-actives are usually identified in plant foods such as fruits, grains, vegetables, and non-food plants such as herbs, spices, aromatics, peels, and sometimes in plant waste materials from factories. Even though they are not recognized as essential components, different experiments showed that they have a role in enhancing human wellness and health.

A plethora of bioactive compounds extracted from natural sources by using supercritical CO_2 for extraction have been shown to exhibit considerable antimicrobial, antioxidant, antiseptic, antibacterial, antifungal, antiviral, anti-inflammatory, antitumor, anti-obesity, anticholinesterase, phagocytotic, and therapeutic activity [207–218]. In the food industry, they can also play a key role as functional food constituents, such as coloring, flavoring, preserving food additives, fragrances, authenticity indices, and biomarkers in metabolomic pathways. The synergism in some of the bioactive compounds enhances their bioactivity as discussed in Chapter 4. As the extraction process is the leading method to recover bioactive compounds, supercritical CO_2 extraction has been reported as the best modern technology to extract bioactive compounds sustainably and safely [219].

5.2.4.6 Carbon dioxide (CO_2): effects on extraction of bioactive compounds

Gaseous carbon dioxide, CO_2, has been extensively used as a reliable supercritical fluid in fields covering foods, pharmacy products, nutraceuticals, cosmetics, fragrances, and folk medicine for extraction purposes due to its critical properties (T_c = 31.1°C, P_c = 73.8 bar), which makes it an inert solvent for extracting bioactive compounds specially those very sensitive to heat [207]. The solvent's unique thermophysical properties have made it the most desirable chemical in the separation of antioxidants, fatty acids, flavors, fragrances, pigments, and essential oils from plants and animal sources. Thus, the use of supercritical CO_2 extraction has numerous benefits: (1) It is a cheap reagent and readily available around the world in large quantities by possessing high degree of purity. (2) It is credited as an environmentally friendly substitute of liquid organic solvents and is designated as a safe solvent to be used in foods and pharmaceutical processes by different regulatory organizations such as the US Food and Drug Administration (FDA), the European Food Safety Authority (EFSA), and the Therapeutic Good Administration of Australia (TGA), which organizes solvents into classes according with their potential toxicity and imposes limits on solvent concentrations in final products made for human consumption [208, 209]. (3) A further advantage of SC-CO_2 solvent is its ability to produce extracts free of solvent residue because CO_2 is a gas at ambient temperature and pressure; so it allows the use of a low amount of solvent and minimizes the thermal damage of the product's bioactive compounds due to its low critical temperature. (4) The CO_2 used in SCFE processes exciting features as a solvent because it has high diffusivity; is safe for human health and the environment; is reusable, inert, nontoxic, noncorrosive, odorless, tasteless, neutral, and nonflammable; and is fully accepted as a food-grade solvent with FDA, EFSA, and GRAS status [210].

According to several researchers and technology users, an advantage of using CO_2 in supercritical fluid extraction processes is that at ambient conditions anywhere, it can readily penetrate the cell wall of plants and dissolve a large number of materials designated as targeted extracts.

Nonetheless, CO_2 at liquefied conditions does not have a straight extraction ability with some substances and its performance will depend on the occurrence of specific chemical groups in the raw material structure, as well as on the polarity and molecular weight of the target's solute [211, 212].

Therefore, to overcome some of CO_2 throwbacks, the extract's composition for volatile or nonvolatile molecules can be fixed easily by modifying the pressure and temperature conditions in the reaction's reactor. Because CO_2 is a gaseous substance at ambient conditions, it makes the analyte recovery quite simple and provides solvent-free analytes. But this property has a limited solubility activity toward certain compounds because it is regarded as a nonpolar solvent and solubility of this kind of substances was analyzed (in Chapter 2).

Several studies had showed that hydrocarbons and other organic substances with relatively low polarity and low molecular weights (MWs) under 250 g/mol exhibit excellent solubility in SC-CO_2 extraction. In this regard, aldehydes, esters, ethers, ketones, lactones, and epoxides are easily extracted in SC-CO_2, so the process can be carried out at lower pressures (75–100 bar; 74.0194–98.69233 atm, respectively). Also, moderately polar components with MW in the range of 250–400 g/mol, such as benzene derivatives, substituted sesquiterpenes, or oleic acid, presenting moderate solubility in SC-CO_2, are targets for extraction, but an increase in energy consumption is required due to higher pressure and is necessary for their extraction to compensate its moderate solubility.

For highly polar substances (MW over 400 g/mol) containing carboxylic acids among three or more hydroxyl groups in their chain structure (e.g., sugars, tannins, proteins, waxes, carotenoids, or pesticides), they are almost insoluble in SC-CO_2 [212]. Therefore, to perform an effective extraction of heavy polar targeted compound(s) like those referred above, a little volume of a polar co-solvent such as methanol, ethanol, and/or water is generally recommended regarding solvation activity of the non-polar solvent used as supercritical fluid (SC-CO_2) to be more liquid than gas in the media by improving the solute solubility in the biphasic solution (liquid/gas) enhanced or facilitated by the presence of a co-solvent [213, 214].

5.2.4.7 Co-solvent effects: colligative properties of CO_2 and modifiers

By definition, a co-solvent is a liquid organic solvent that, when added at a specific proportion to CO_2 fluid contained in a sealed reactor, may dissolve with the supercritical fluid and will retain by itself its characteristic solvent power toward the targeted compound groups [215].

As previously discussed, CO_2 has been the most frequently used supercritical fluid in various industry extraction processes, but its low polarity is accompanied by concerns and restricts its use for the extraction of lipophilic and polar substances. Therefore, it is a less effective solvent in, for example, the extraction of polar phytochemicals embedded in the cell walls.

To overcome this problem, polar co-solvents such as ethanol, methanol, water, acetic acid, formic acid, and many other polar or nonpolar co-solvents can be used for a polar substance extraction to improve the solvation power of SC-CO_2, by enhancing its affinity for poorly soluble solutes (e.g., alkaloids, phenolics, and glycosidic compounds), increasing their solubility and the extraction yield and thus improving the recovery of bioactive compounds [216, 217]. Among the foremost used co-solvents, methanol and ethanol are most frequently used at concentrations below 10% of the CO_2 employed for a co-solvent-assisted extraction [218]. Nonetheless, for compounds with human or animal application, ethanol became the co-solvent of choice due to its broad use in the medicine and liquor industry because of the less toxicity when compared with methanol [219]. Experimentally, the best co-solvent that could be used for a specific extraction must be determined through preliminary assays by using a little sample amount, have a clear knowledge on the physical state of sample used (e.g., gel, solid, liquid), the target chemical compound(s) to extract, and whether the co-solvent in mind could have collateral reaction activities to be considered as adverse [220].

Different investigations had showed that the addition of co-solvent can affect not only the total extraction yield but also the bioactive properties of the extracts, such as the total phenolics extraction yield, and the antioxidant and anti-inflammatory activities of certain extracts [221–223].

Thus, several research works agree that the percentage and the type of co-solvent used in supercritical fluid extraction are parameters having a significant impact on the whole process. Hence, both are crucial factors that together affect the extract solubility of the targeted compounds in addition to the solvent's fluidity inside the cell matrices, improving the product yield.

For example, the addition of large volumes of co-solvent could modify the critical parameters of the solution, which would be reflected on a substantial reduction in selectivity by the extractant fluid [224]. Consequently, there are three different proposed ways through which a co-solvent can be used in an SCF process: (1) as a mixed fluid added in the pumping system; (2) by injecting the co-solvent as a liquid together with samples in the reactor or vessel before extraction; and (3) contained in a separated cylinder tank connected to the pre-modified CO_2 tank, although this method is thought to be expensive and rarely used [225]. On this regard, several scientists have studied the effect of co-solvent-induced solubility on SC-CO_2 extraction, and their results have various implications with an elegant description made by Taylor [222].

For example, the carotenoid chemical composition of pumpkin after its SC-CO_2 extraction with and without co-solvent showed that due to the presence of a co-solvent, an increase in the total carotenoids yield from the pumpkin's sample was observed contrary to a significant decrease in yield by the absence of it. Similar studies performed by Shi et al. showed that by increasing the polarity of CO_2 with ethanol, it was possible to increase the total carotenoid yield by 1.8 times compared in its absence [217]. Several research groups had studied the impacts of ethanol related with its concentration and the extractor pressure on the yield of phenolic compounds, the total extract yield, its antioxidant, and anti-inflammatory activities by using supercritical extraction of *Lupinus rivularis* stalks [218].

Other studies correlating extraction process and structure–activity relationship showed that the highest extract yield, targeted phenolics yield, and antioxidant activity were achieved with the low use of co-solvent, while anti-inflammatory activity was improved after a great addition of ethanol co-solvent. A group evaluated the effects on the extraction of phenolic compounds by using two different co-solvents (ethanol and ethyl acetate) together with modifications on the pressure of SC-CO_2, and their results revealed that the extraction process is largely favored by the addition of ethanol as a co-solvent than those experiments obtained using ethyl acetate for example [223]. Moreover, it is thought that the solvating power of a supercritical fluid at a given temperature and pressure would depend on its density, ρ [221]. From these and other arguments on the solubility power of a solute under the presence of fluid CO_2; on the other hand, it will depend also on the intermolecular interactions between solvent and solute in the same phase [224]. On this reasoning, the chemical solvating power of supercritical CO_2 as a fluid can be determined in terms of the solubility parameter S. This parameter is defined as the square root of the cohesive energies of each component in the extract solution, which leads to the system density and is used extensively in interpreting the thermodynamic properties of a solution, by taking into consideration its physical state, and is given by

$$S = C^{1/2} \tag{5.2}$$

with

$$C = (\Delta H - RT)/V \tag{5.3}$$

where ΔH is the heat of vaporization (kJ/mol, cal/g), V is the molar volume of the fluid, C is the cohesive energy, R is the gas constant, and T is the experimental temperature (K). If we analyze a supercritical fluid at its critical temperature, the heat of vaporization, ΔH, is zero, so then the result of Eq. 5.3 leads to a negative value of C.

Thus, the equation had been modified, and Giddings et al. proposed an equation based on studies using liquid chromatography by applying the van der Waals equation of state [225]. In that case, the solubility parameter, S, of a SFC is given by

$$S = 1.25 P_C^{1/2} \frac{\rho_{r,SC}}{\rho_{r,liq}} \tag{5.4}$$

where $\rho_{r,SCF}$ is the reduced density of the substance in its supercritical state and $\rho_{r,liq}$ is its reduced density in the liquid phase at the substance normal boiling point.

Therefore, the solubility parameter, S, for a given solvent can be obtained even if the fluid density and critical pressure are known, as these properties are unique for a given substance. Furthermore, when the solubility parameters of both the solute and the solvent are similar, an appreciable solubility of the solute will generally be observed in the given solvent although their polarities differ in some extension. The solubility parameters of some co-solvents used in supercritical fluid extraction are given in Table 5.5.

5.2.4.8 Effects of experimental conditions on lipid extraction by SFE

The extraction of bioproducts possessing added economical value by using SFE (e.g., SFE-CO_2) is well known at both academic and industrial levels and has been the subject of numerous publications. The industrial interest in the subject has gained track in the last two decades because the large-scale application of SFE technologies has been well known since the late 1970s for the decaffeination of coffee and tea natural products. Currently, numerous applications of SFE-CO_2 are undergoing for the extraction of natural substances such as edible oils, fatty acids, light flavor molecules, fragrances, and pigments from various plants as well as from microorganisms (like algae), and by-products must continue to be the main theme of further research activities. The increased use of SFE-CO_2 technology for the extraction of natural substances is a select theme of research and broadly described in books, papers, and reviews [222–225].

Table 5.5 Critical properties of co-solvents most used in SFE

Modifier	T_c (°C)	P_c (bar)	ρ_c (g/cm^3)	*WM* (g/mol)
Water	347.1	217.60	0.322	18.01
Dimethyl ether*	127.2	53.56	0.272	74.12
Ethanol	243.0	63.0	0.280	46.07
Methanol	239.4	79.9	0.275	32.04
1-Propanol	263.5	51.0	0.273	60.10
2-Propanol	235.1	47.0	0.274	60.09
1-Hexanol	336.8	40.0		102.18
2-Methoxy ethanol	302	52.2		76.10
Acetonitrile	275	47.7		41.05
Dichloromethane	237	60.0	0.440	84.93
Chloroform	263.2	54.2	0.498	119.38
Tetrahydrofuran	267.0	51.2		72.11
Dimethylsulfoxide	465.0	56.3		78.13
1-4 Dioxane	314	51.4		88.11
Propylene carbonate	352.0			102.09
N,N-Dimethyl acetamide	384			87.12
Formic acid	307			46.02
Carbon sulfide	279	78.0		76.13

Adapted and modified from Ref. [222].

For example, essential oils (EOs) were traditionally extracted from plant seeds, roots, bark, flowers, and leaves through hydro-distillation, thermal degradation, hydrolysis, and by taking advantages of the solubility of some compounds in water that may alter the flavors of the desired chemical compounds of the final product. Nowadays, it is thought that by employing SFE-CO_2 as an efficient alternative technique to avoid these and other undesirable experimental problems, the optimum operating parameters for the extraction of essential oils by the SFE-CO_2 method are mainly pressure and temperature in the range of 90–250 bar and 40 to 50°C, respectively. Thus, Conde-Hernandez et al. [30] published their work on the utilization of the SFE-CO_2 technique for the isolation of EO from rosemary. The parameters considered were temperature (40 and 50°C) and pressure (10.34 and 17.24 MPa) at two levels, and the maximum EO recovery was between 1.41 and 2.53 g EO per 100 g of dry rosemary (% w/w) for the conditions selected.

5.2.4.9 SFE with dimethyl ether as co-solvent for lipid extraction

The extraction of lipids from soya beans at various solvent ratios was investigated using SC-CO_2 and dimethyl ether (DME) as co-solvent and reported as an update regarding the effect of a pure supercritical fluid and a mix of this with a co-solvent under critical conditions. To do that, both temperature and pressure were set up to 60°C and 20 MPa, respectively, with the extraction stopped when the extracted amount was understood to be constant.

It was determined in the early stage of the extraction that when the amount of solvent flowed was less than 0.1 kg, the amount extracted per solvent (the inclination in Fig. 5.10) was very high, especially at a ratio of 9:1, where the yield reached 18.5% at a solvent volume of 0.094 kg, which is 91.6% of the total amount extracted.

Thereafter, as the solvent flowed constantly, the amount extracted per solvent decreased at the same CO_2:DME ratio. Thus, the Fig. 5.10 shows the yield product that comprises the amount of extracted lipids based on SC-CO_2 at 60 °C and 20 MPa under the effects of co-solvent dimethyl ether, DME from soya bean at different mixing ratios. Dry sample weight calculations were found to be 12.9% in the rate of extraction setup, by taking parameters of solvent consumption to be approximately 0.1 kg with an extraction rate of 5.85% for a ratio of 0, 6.49% for 29:1, 10.1% for 14:1, and 18.3% for 9:1, respectively, which suggests that the solubility of the extracted lipids of this natural material (soya bean) in each solvent is almost different. Therefore, under all experimental conditions, it was sought that when the solvent mixture exceeded 0.4 kg, no changes were observed in the extraction volume, so the time required to flow above 0.4 kg was determined to be 116.1 to 118.8 min (~2 h) to complete the extraction [226].

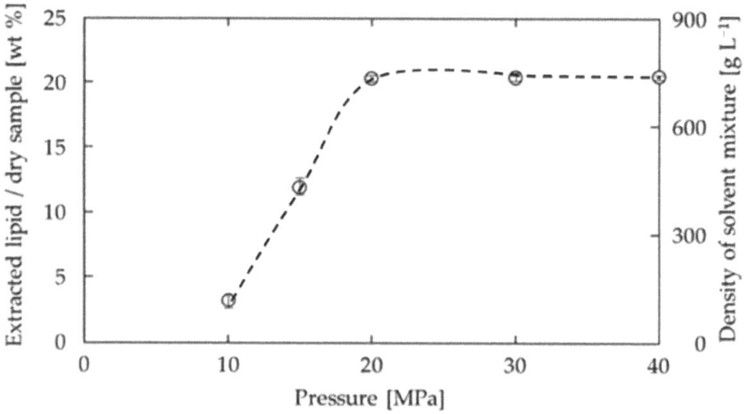

Figure 5.10 SC-CO_2 at 60°C and 20 MPa under the effects of co-solvent dimethyl ether (DME) on lipid extraction from soya bean at different mixing ratios. Adapted from Ref. [226].

The effects of the pressure of the solvent mixture on the amount of extracted lipids are shown in Fig. 5.11. The results show that by setting the pressure between 10 MPa and 15 MPa, the amount of lipids extracted was low, and that the maximum extraction was obtained when pressure reaches 20 MPa, but the experiment also shows that above this pressure, the lipid yield does not increase. Similar results were obtained with extractions carried out by utilizing pure SC-CO_2 at 60 °C and 20 MPa, interestingly, these yields decreased to 15.8%, and 19.1%, with the experiment carried out at 30 MPa, and 40 MPa, respectively. Therefore, when dried soya bean extraction is performed with SC-CO_2 at 20 MPa using DME as co-solvent, the process yielded more lipids than the extraction at 40 MPa.

The experimental data showed that there was no difference in the yield of lipid extracted and that there were not significant differences in the amounts extracted per solvent, or no changes were observed over the time, as shown in Fig. 5.12. At this point and as observed in Fig. 5.12, the SC-CO_2 to DME ratio is the same 9:1 and the temperature is also 60°C.

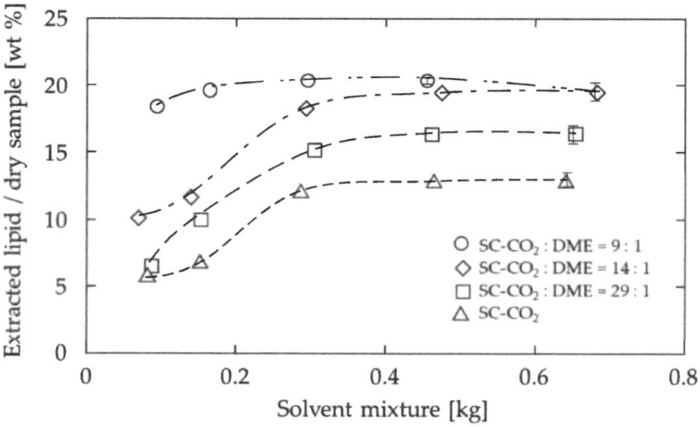

Figure 5.11 The effect of pressure on the yield of lipid extraction from soya bean through supercritical fluid CO_2 with co-solvent DME (9:1) mixed at 60°C. The dashed curve is the density of solvent mixture. Adapted from Ref. [226].

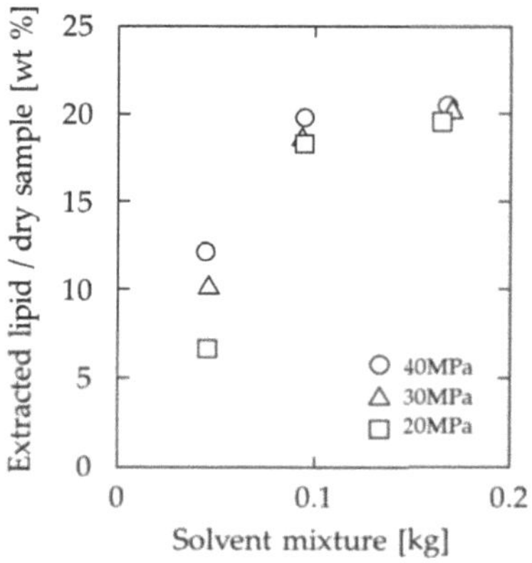

Figure 5.12 Extraction yield of lipids from soya beans in the early stages of extraction at 20, 30, 40 MPa by SC-CO_2 : DME (9:1) at 60°C. Ref. [226].

Figure 5.13 shows, thereafter, that at the initial stage, the higher the pressure, the higher the amount of lipids extracted. Thus, when the solvent mixture flows approximately at 0.044 kg/min, the lipid yields recovered were 6.7 wt%, 10.4 wt%, and 12.2% at 20, 30, and 40 MPa, respectively.

On the other hand, the temperature effects of using the SC-CO_2–DME mixture at 9:1 ratio and 20 MPa are shown in Fig. 5.13.

When this parameter was set up at 80°C, it significantly reduced the amount of extracted lipids compared to that recovered at 40°C and 60°C, respectively.

Therefore, as discussed until here, it is not difficult to hypothesize that supercritical fluid extraction using CO_2:DME mix can be considered a technology more temperature dependent due to the thermodynamic dimension that determine the efficacy of CO_2 as solvent is it's density and lipids vapor pressure shows they are molecules with high temperature dependence for their physical chemistry interaction with a solvent. Thus, an analysis of the collected data shows an increase in the total amount of lipids extracted from the soya bean with an increasing proportion of co-solvent DME in the fluidic mixture, which can be attributed to the strong intermolecular interaction of DME co-solvent toward CO_2, with both in the liquid phase.

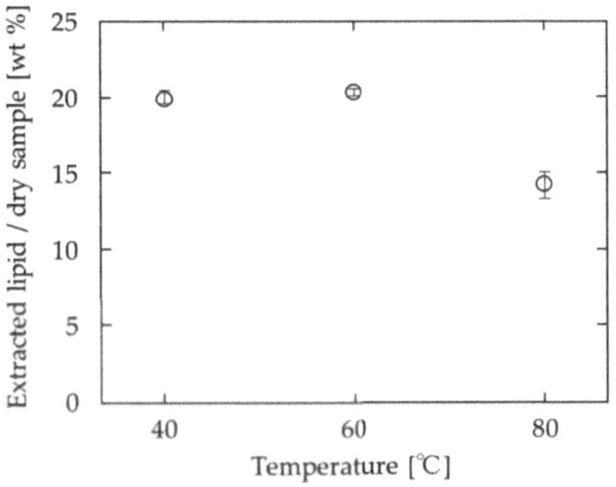

Figure 5.13 Effects of temperature on lipid extraction from soya bean by SC-CO_2:DME (9:1 ratio) at 20 MPa. Adapted from Ref. [226].

This interpretation means that a mix of CO_2 and DME in the 14:1 ratio is the most efficient supercritical solvent for the extraction of a large amount of lipids in this type of soya beans, however and as discussed above, CO_2:DME solution in the 9:1 ratio, is the most desirable solvent/co-solvent mix that improves lipids solubility together with great impact on extraction time reduction.

In this specific case, the 14:1 ration results between CO_2 and DME can be also extrapolated to analyze the direct effect that other mixture ratios can has on the solvent pressure and its direct effect on lipids yield extraction for other type of soybeans or specific crops that shows to be directly related to the density of the supercritical mixture under specific or selected conditions. This correlation was evident by the solid curve obtained in Fig. 5.11, which shows how the lipid extraction yield is pressure dependent regarding the density of the solvent mixture formed by CO_2 with DME at 9:1 ratio, 60°C and at either 20 or 40 MPa of pressure.

From the experimental data, it is observed that the density of the CO_2/DME mixture increases significantly in the interval of 10–20 MPa, but above 20 MPa, the increase in that density is too small to promote any change on the lipid's extraction yield. Thus, the pressure dependence on the mixture density shows a similar trend to that observed on the pressure dependence of the yield of lipid extracted. Nonetheless, the maximum amount of lipid extracted was almost the same at and above 20 MPa, which suggests that a SC-CO_2 system operated at a pressure of 20 MPa is enough to obtain an efficient solvent–lipid interaction, leading to the best extraction of this compound from soya beans in a short time.

Similarly, the temperature dependence on the lipid's extraction volume showed its importance attributed to substantial changes in the density of SC-CO_2 directly related with the total lipid solubility in the supercritical fluid at different temperature and pressure. Therefore, it is not incorrect to say herein that the solubility of a solute in a solvent is strongly influenced by the density of the solvent and the saturation of the vapor pressure of the solute in the extracted material [226]. For example, by increasing the temperature and getting a constant pressure, the data show an increase in soya bean lipid solubility (see Fig. 5.10), because at high temperatures, they evaporate more easily, and by maintaining a fixed thermodynamic parameter (pressure), the solute achieves favorable conditions for its saturation point where they would quickly dissolve in the hot supercritical fluid. At this point, however, the density of SC-CO_2 decreases, which could weaken the solute/solvent interaction between lipids and carbon dioxide molecules, so then, a deep decrease in solubility should be detected.

Therefore, the condition at which solubility is dependent of temperature as a reversed process in supercritical fluid-CO_2 extraction is commonly known as the *crossover point*. At this condition, the crossover point observed for the process described herein was measured as a set of contributions belonging from a decrease in the density of the supercritical solvent mixture, which was more significant than that from the increase in the saturated vapor pressure of lipids detected by an increase from 60 to 80°C at constant pressure.

The data shows however, that a crossover point exists in the mixture formed by the supercritical fluid-CO_2 and DME, as well as from other extraction carried out with pure SC-CO_2. From these results the analysis suggest that lipids obtained from dry grains among soya beans through CO_2-DME mixture or pure CO_2:

- No differences were found in the yield of lipid extraction if the pressure exceeds 20 MPa.
- The process efficiency is temperature dependent with best results occurring at 40°C and 60°C at the same pressure, but a decrease is observed at 80°C.
- The presence of DME as co-solvent increases the amount of lipids dissolved in the supercritical fluid mixture.

5.3 Outlook on Carbon Dioxide for Green Extraction Technology

Labeling carbon dioxide as a "green solvent" may be misleading or lead to confusion. It is because the carbon dioxide used in supercritical fluid extraction is mostly collected as a by-product in chemical manufacturing (e.g., NH_3 and ethylene oxide) or by trapping methods such as post-combustion, and in petroleum-reforming processes. If we follow the completion of the above-mentioned SFE process, the CO_2 reverts to a gas at atmospheric conditions, and then the stored CO_2 (bottled or buried) ultimately makes its way back into the atmosphere. Therefore, the net change in CO_2 levels is zero. Furthermore, using a gas/extract exchange chamber, the "clean" CO_2 (i.e., as a gas solvent and after separated from the extract) can be recirculated for continued reuse, and this can happen, at least theoretically, over an indefinite time scale. But such a recycling chamber needs to be added to the overall establishment costs of the operation system and the ongoing additional running costs of periodical maintenance. Additional advantages encountered with CO_2 SFE methods are the low critical temperature that characterizes the physical properties of this molecule, meaning that some thermolabile compounds can retain their functional properties without chemical damage, nor medical activities suppressed during the extraction process. However, a major drawback of the use of carbon dioxide as a solvent in its SCF state is that it is only suitable for CO_2-philic molecules, by rendering its best efficiencies only with a small number of nonpolar molecules, because CO_2 is intrinsically a nonpolar compound. Fortunately, this physical chemistry limitation can be overcome by the addition of a more polar co-solvent as discussed in the lipid extraction from soya beans. Furthermore, a supercritical fluid can, in fact, be used as an anti-solvent, by taking the advantage that a plant material can be dissolved initially in an organic solvent and subsequently be injected into an SCF phase (usually supercritical CO_2), where the miscible organic solvent (e.g., DME) and liquified CO_2 interact in a sealed camber, which together reduces the organic solvent phase and the dissolved solutes can precipitate, leaving the extract to be examined as a chemical product or for its bioactivity property. Nonetheless, the anti-solvent SFE methods have been addressed elsewhere and will not be discussed further in this chapter [226–230].

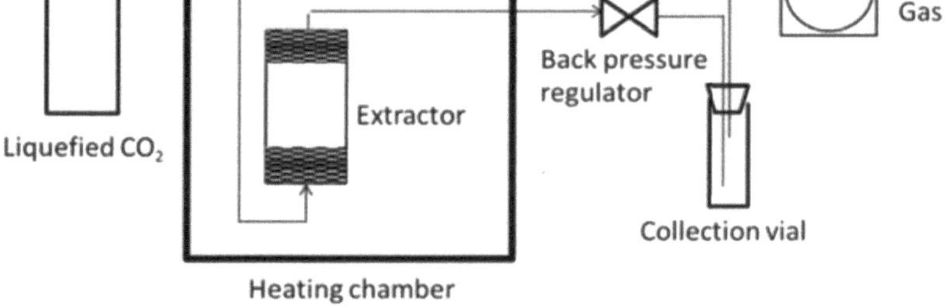

Figure 5.14 Schematic diagram of supercritical fluid CO_2 system for lipid extraction using DME as co-solvent. Adapted from Ref. [226].

In general, a supercritical fluid extraction system (Fig. 5.14) can be divided into three areas of application, such as an academic laboratory, a pilot plant, and at industrial scale. For a laboratory-scale system, it aims to produce from milligrams to a few grams of extract by using small volume reactors (e.g., 50 to 300 mL) [231]. For an increased order of needs where kilograms of extracted material are desired to be processed, the extraction system will comprise a large reactor where large volumes will be required to accommodate from 10 to several hundreds of liters [232]. To satisfy these needs, the basic components of a typical CO_2 supercritical fluid system consist of a CO_2 pump or a compressor, a modifier fed from a pump as an organic co-solvent or polar entrainer (including water) when required, an extraction vessel or reactor, and a fractionation/collector vessel to recover the extract. To achieve the selective extraction of a target compound or a chemical family, a sequential batch of separators formed by multiple reactors connected into a series has been introduced as an efficient alternative that makes supercritical fluid extraction a more profitable technique [233]. Other possible way to achieve this goal is by appropriately varying the pressure with different frequency, which will create successive fractionation of various classes of compounds for specific purposes [234].

Some examples for the latter suggestion (at industrial scale) includes the physical optimization of the reactor vessel (stainless steel, fiberglass internal cover) in an aims to enhance extraction of phenolic compounds, prevent corrosion due to acid activity of citric fruits, improves fractionation for vanilla oleoresins, ginsenosides, and polysaccharides from Panax ginseng as well as extraction of compounds used as anesthetics or cognitive medicines (e.g., alkaloids, tocopherols, tocotrienols, and γ-oryzanol) extracted from rice bran as batch stages proposed in other chapters [233–236]. Finally, the yield of the extract is an important outcome to be measured and abundant research data suggest that the yield of target compounds can be improved mainly by sequential extraction (e.g., multiple extraction vessels connected in series) of natural products using different solvents or even mixtures of solvents at each subsequent stage [237].

Chapter 6

Semi-Critical Assisted Extraction: Insights in the Integration of Empiric and Advanced Extraction Technologies

6.1 Introduction

The complexity in chemical composition alongside the diverse difficulties encountered in the extraction and/or separation presented by crop plants, animal tissue, agriculture bioproducts, pharmaceutical products, and algae biomass poses significant challenges for a rapid analytical characterization of their biotechnological added value.

The extraction and identification of amino acids, lipids, antioxidants, and proteins from several sources had been performed mainly through empirical (maceration, percolation), Soxhlet (temperature and time dependent), and recently sub- and supercritical (temperature, pressure, CO_2 dependent) methods.

Several designs of glassware equipment have been proposed in the past to extract chemical substances from solid, liquid, or gel states with the chemical extraction carried out by two possible processes: (1) continuous and (2) discontinuous. Thus, the extraction and concentration of substances from samples such as water contaminants, plant materials, and biological origin medicines are carried out in two different apparatus, one for extraction and

Semi-Critical Assisted Extraction: Applications and Commercialization in Biotechnology, Food, and Pharmacy
Tulio Chavez-Gil

ISBN 978-981-4968-36-2 (Hardcover), 978-1-003-29124-4 (eBook)
www.jennystanford.com

another for targeted molecule(s) separation or occasionally, using a combined extractor/separator technology. In these cases, the chemicals to be extracted are dissolved by means of an extractant in its either steam or liquid phase throughout a continuous extractor apparatus, with the sample confined in an extraction space containing the material having the extractable component. Depending on the technology choice, the extracted substance is dissolved with the solvent within the extraction space and can be directed as an extract/solvent mixture to a reaction flask in a cyclic process in which the extract will be exposed to a solvent boiling temperature for extended periods of time.

This type of extraction method has been used in the food, pharmaceutical, and folk medicine industry for the last 120 years using a universal design, the so-called "Soxhlet extractor," which allows for a convenient automatic cycle events for returning the extract toward a reaction flask using a siphoning system described by the German Institute of Standard 12602, which occurs when the condensed solvent level saturated with the content within the extraction space is reached. In the system, when steam from the boiling solvent reaches the extraction space in the conventional Soxhlet extractor vessel, this causes the condensate to still be hot when it drips through the stationary sample, wherein the extracted substance is then siphoned toward the reaction flask where the solvent will be boiling indefinitely [238].

This extraction method is well suited for processes where the sample density is understood to be greater than the solvent density, as well as for molecules with high molecular weight. But the method shows poor efficiency with respect to the quality and purity of the extracted product(s) for samples having densities lower than the solvent density or for biological substances suspected to be thermolabile, such as amino acids, enzymes, proteins, edible oils, etc. A similar extraction method known as *enriched extraction* has been implemented using a "Gregar extractor," developed at the Argonne National Laboratory, Chemical Engineering Division, where the continuous extraction principles previously explained are evident in the instrument design and its related methods. However, one disadvantage found in the conventional continuous extraction apparatus lies in the difficulty to perform the extraction through the supply of a fluid in two directions at the same time and the lack of monitoring the fluid temperature before it reaches the sample inside the extractor. Therefore, when more than one solvent is needed for extracting a substance of interest, the ideal process to reduce the working time would be to supply a mixture of solvents in a single step considering that more than one fluid will condensate over the sample if the boiling point of the mixture components is too close, so then, additional work would be necessary to separate the extract contained in the mixture.

6.2 Thermodynamics of Semi-Critical Extraction

Subcritical, critical, and supercritical extraction methods had been described to require the use of practically the same instrumentation that can support extreme thermodynamic conditions (critical temperature, T_C, and critical pressure, P_C) as key conditions to perform successful experiments carried out by utilizing similar solvents to achieve profitable efficiency and chemical quality of valuable compounds with pharmaceutical, food, agriculture, and health wellness commercialization. On the other hand, *semi-critical extraction* (SmCE) can be performed in an instrument designed with intermediate thermophysical particularities in that a solvent's temperature and pressure are allowed to be raised above standard, but far below critical conditions, by forming a constant fluid biphasic (liquid/vapor), due to smooth changes in one of these two physical variables (T or P).

Semi-critical assisted extraction is then easily carried out in the Ch-G extractor [238] after key physical and engineering designs were combined to its construction as a compact extractor/separator vessel (initially in glass) that works at isothermal or isobaric as well as under semi-critical conditions, allowing the exposition of different raw material and cell matrices to be shacking by a constant and dynamic fluid aiming to improve extraction or separation of valuable molecules such as amino acids, proteins, vitamins, antioxidants, medicines, algae biofuels, crop oils, or contaminants (e.g., pesticides, polyaromatic hydrocarbons, PAH's), within others. A special feature of the SmCE method is characterized by the high quality of recovered organic and inorganic chemical composition as well as molecules possessing high bioenergy content [239].

Previous glassware extractors have been designed to equilibrate the density difference between the solvent and the solute, which requires additional work to maintain the level of each component by avoiding the fluid to flow in the opposite direction within the apparatus. In the Ch-G compact vessel the biphasic (solute/solvent) pressure is leveled through the manual operation of special features (stopcocks A, B and a sphere) designed to adjust the system hydraulic pressure between the solvent and the extract.

After reaching this goal, some extracting apparatuses have additional components and features that allow an operator to equilibrate key physical differences with some degree of operational risks. Thus, such extractor technologies are subjected to special operational care directly related to their design, which includes the assembly sequence of the equipment to reach the desired equilibrium between the extract and the solvent volume, through an either external or internal embed component or a piece that allows the balance of the hydraulic pressure inside the extractor vessel and that satisfy the proportional relationship:

$$S_h \times S_\rho = sol_h = sol_\rho \tag{6.1}$$

where S = sample, h = height, ρ = density, and sol = solvent.

Nonetheless, a technician working with an equipment designed to satisfy Eq. 6.1 will not always be mindful of reaching the ideal level of the relationship *sample: solvent* that satisfies the balance between them according with the designed parameters that operate over the whole parts of the equipment. In these cases, it would be desirable to have a more conventional extraction equipment possessing a related method that simplifies the process with a design that allows the operator to balance both the level of fluids (mainly liquids) and the hydraulic pressure in a quick, simple, and safe manner. To that, when pure or a mixture of solvents is heated in a reaction flask, the steam generated creates a linear or crossed flow inside the conducting container (generally a cylinder) or over a mechanical interfering part (e.g., a marble sphere), by exhibiting a complex pattern according to a thermodynamical analysis.

It is thought that in its liquid and gas phases, a molecule shares the same two intensive properties (density, d, and viscosity, ρ), which are directly related to the fluidity and velocity of any phase inside a transporting system. However, the gas phase is more fluid than its original liquid when analyzed in a transporting phenomenon, and thus, its viscosity, which is defined as the resistance that a part of the fluid shows to the displacement of the other one, controls the process. Therefore, viscosity, ρ, is produced by a cutting effect of a layer of fluid in one phase when it is displacing over the other and is completely different than the so-called *intermolecular attraction*.

By assuming that a liquid stratifies in molecular planes, a plane's area is defined as A, and the interplanar distance between phases can be defined as dy. In addition, if we assume that each plane moves toward the right side with velocities v_1, v_2, etc., where each value is greater than its predecessor by an increment of dy, then a flow occurring according to this description is called *laminar* and is different than the *turbulent,* where the plane's parallelism is not observed. Thus, in the laminar flow, the force required to maintain a stationary velocity difference dv between two parallel planes is directly proportional to the magnitudes of A and dv and is inversely proportional to the plane's distance dy. Therefore,

$$f = \eta A \frac{dv}{dy} \tag{6.2}$$

where f is fluid force, η is fluid viscosity coefficient, or simply fluid viscosity, while the amount dv/dy in Eq, 6.2 refers to the cut velocity v_c, and the relationship f/A, force per unit of area, is called the cutting force F. Thus, in terms of v_c and F, Eq. 6.2 can be written as:

$$\eta = \frac{F}{A} \tag{6.3}$$

In this way, Eqs. 6.2 and 6.3 could be taken as expressions that define η and the practical application of these physical properties depends on the validity of a series of experimental assumptions, especially when the solvent flow is described as laminar. Therefore, when designing a fluid transportation system, the onset conditions for turbulent flow depend on the magnitude of a certain combination of experimental variables pertaining to a pure number called the Reynolds number, R_e [238]. Thus, for a flow through a large, cylindrical, and linear pipe, this number is obtained from the equation:

$$R_e = rv\frac{\rho}{\eta} \tag{6.4}$$

where ρ is the fluid density, r is the pipe radio, and η is the viscosity coefficient that herein correspond to a laminar biphasic fluid. It has been found empirically that a laminar flow is always obtained in the same pipe when R_e is less than 10^3, by virtue of the magnitude of any individual variable such as r, v, ρ, η.

Nonetheless, the laminar flow of a solvent fluid depends on the regularity of the wall surface as well as on the entrance of the pipe, among the internal length L' of the transition region, since it is very important that this transversal section needs to be very small in comparison to the pipe length, in a relationship:

$$L' = \frac{1}{4} R_e r \quad (6.5)$$

where r is the pipe radius and R_e is the Reynolds number. It is inferred from this relationship that the pipe must be tight to obtain a laminar flow and the Reynolds number could be considerably reduced if the pipe is substantially curved [238]. According to the prototype depicted in Fig. 6.1, the fluid's temperature and pressure can be continuously monitored in an extraction or separation, regardless of the physical state of the material used as sample (solid, liquid, or gel).

6.3 Advantages of SmCE Method

In semi-critical solvent extraction, the extracted substance is concentrated within the space of the extractor vessel (Fig. 6.1, pipe 1) with the possibility to be (1) drained toward the reaction flask in a continuous extraction process, (2) directed toward a separation flask continuously or step by step, with the collected fractions used for chemical or physical analysis, and (3) separated as the final product.

The SmCE apparatus has been designed following basic thermodynamic principles of fluid relationships such as pressure and temperature occurring in closed pipes with different diameter and length, to makes the extraction process simpler, safer, and cost effective.

A special feature of this innovation allows the solvent to flow initially upward through a short chamber formed by a pipe of a reduced diameter which is fused in its upper side to a large pipe. This design transforms the downside of the fused pipe into the first pressure exchange chamber because its upper side is obstructed by a sphere that rests freely over a concave surface inside the crossing pipe and divide the large pipe in two sections, a below exchange pressure chamber before the sphere and a large wide pipe after the sphere.

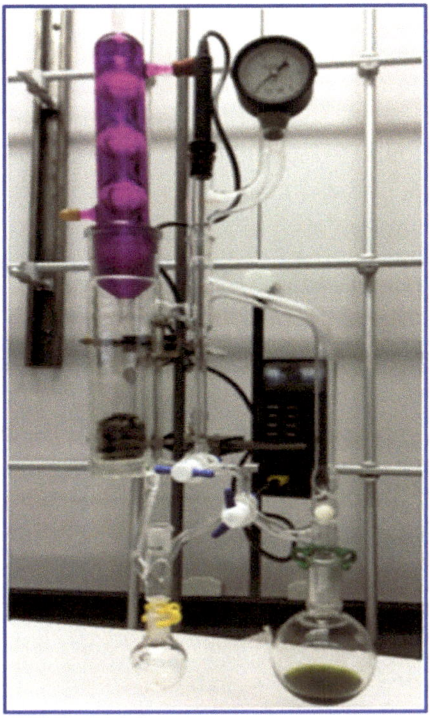

Figure 6.1 Ensemble of a Ch-G glassware prototype for semi-critical extraction fitted with an analog device and a digital device (gauge and thermocouple) for determining the in situ thermodynamic properties. Personal picture by author.

The below exchange pressure chamber in turn, has a reduced lateral connection port that communicates with a pressure control valve (B). According to another aspect of the innovation, some particles of the fluid can be concentrated in the space having a greater diameter, while other particles can travel through a reduced section pipe by exerting upward pressure below the surface of the sphere on its middle plane, which when striking the sphere causes the speed of the fluid to diminish or stop completely, wherein at this point the greatest pressure will be reached within the chamber.

Several components for an easy change in pressure are interconnected to allow a non-trained operator to take control of any overflow in the circulation of steam during a continuous process.

For example, the interconnection between different pipes to transport the solvent and the extract can be constructed with different internal diameters (e.g., 1.5–2.0 mm) to provide the

SmCE apparatus with pressure chambers adjusted to routes of fluids circulation, according to the nature of the sample and more important to the density of solvent. Thus, mechanical components can be afforded for changes in the pressure on specific parts because they are interconnected and allow an operator for an easy control on any overflow in the circulation of either steam or liquid extract during a continuous or step-by-step process.

A special aspect of the SmCE apparatus is that it can be made of glass material for an efficient, economic, easy to clean and rinse for continuous or step-by-step extraction capable to operate with small, medium, and large amounts of sample by using a minimum amount of solvent as an extraction means.

The SmCE constructed as glassware is an efficient extraction apparatus, which has provided excellent results for stage-by-stage extraction of organic components from plants and other biological samples susceptible to decomposition when subjected to high temperatures. The glass material construction makes this extracting apparatus a method suitable for an easier extraction of biological oils of seaweed, obtained through azeotropic mixtures of organic solvents in situ [239].

Another advantage of construction in glassware is that the apparatus and the method facilitate the extraction of biological substances such as proteins, amino acids, and enzymes by using pure solvents or azeotropic mixtures of low boiling point. The glass material also allows for the extraction of environmental contaminants dissolved in water as polluting agents, as well as for the separation and purification of organic mixtures to be used in further extraction processes.

6.4 Brief Description of Semi-Critical Extractor Apparatus: The Ch-G

The compact design and special features that characterize the Ch-G vessel had been tested to be especially advantageous in the extraction of organic compounds, phytochemistry material, edible oils from algae, crops, spent coffee ground, as well as of valuable thermochemical susceptible biological substances such as amino acids, proteins, enzymes, and for environmental contaminants of soil, fresh water, and food. In addition, the compact design allows the extractor to be washed or sterilized periodically without the need to be disassembled.

According to the SmCE design (Ch-G), the apparatus can be easily washed to prepare an inert sample to be exposed to a fresh solvent without the need for disarming the equipment as with other commercial extractors.

Therefore, a key advantage in the glass construction of this extractor/separator equipment arises due to two spaces with different capacities: Space 1 is a large vertical vessel which supports the atmospheric pressure on its upper side plus the hydraulic pressure created by the fluid solvent (steam) that achieve this space through an embedded second vertical pipe (2), fused to 1 by means of two connecting cylinders having a reduced section (3) on its upper side and another cylinder of reduced diameter (4) on the lower side, with both connectors positioned on the right side of the equipment with this chamber (1) ending on an open space (stopcock 9) on its upper side available to support a large condenser filled with circulating cooling fluid (Fig. 6.2).

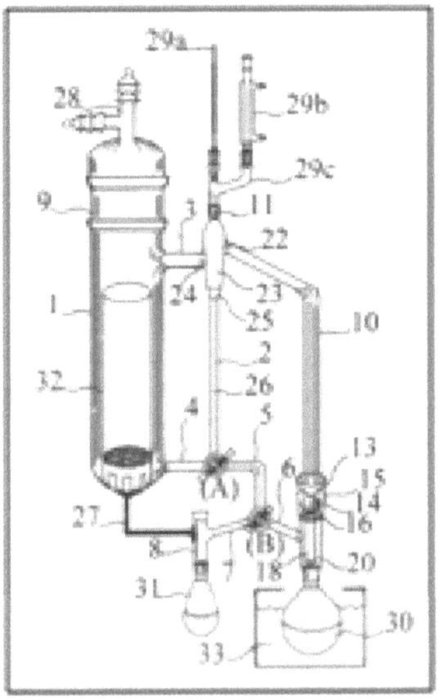

Figure 6.2 Schematic drawing of a semi-critical extractor for assisted solvent extraction. Constructed in glass, thermolabile biological valuable molecules from food, animal tissue, plants, algae, and medicines are safety extracted. Reprinted from Ref. [239].

An advantage offered by the Ch-G extractor is the inclusion of a second parallel pipe 2 as a large and narrow cylinder comprising a special open section 11 on its upper side, designed to accommodate additional peripheral elements that includes but are not limited to analogic (thermometer) as well digital (thermocouple) temperature detector devises, or chillers such as a condenser head 23. This part connects directly with pipe 1 through connector 3 in port 24, followed by a conical section 25 which extends downside with a reduced-section area 26 that is embedded to pipes 4 and 5 throughout a "T" valve A (made in Teflon or glassware stopcock) and embedded to vessel 1, by completing the connection between both main pipes 1 and 2 (Fig. 6.2). The SmCE design at difference to another glassware apparatuses allows the steam of a fluid (liquid or gas) can reach the main extractor vessel (1) in short time due to the differences between the atmospheric pressure in vertical descending direction on pipes 1 and 2; and the hydraulic upper side pressure excerpted by the fluid travelling throughout the pre-cooling pipe 2.

To achieve this goal, the positive pressure of fluid could reach the extractor 1 very fast through connectors 3 or 4 by manipulating the flow control of the valves A and B, which act automatically over the glass sphere 14 with a movement defined by small pressure differences inside extractor pipe 1 and the other parts of the apparatus altogether. Therefore, the apparatus allows a mixed process of extraction/separation in a single step, facilitated by easy manipulation of the two fluid pressure control valves allowing an operator (a) to initially separate from a mixture a highly pure or an azeotrope solvent, or (b) to continue an extraction that requires the use of more than one solvent to ensure the process completion.

In a standard extraction or separation, the sample to be processed is placed directly inside the extractor 1 vessel, or inside a container like a cellulose or a porcelain glass-sintered thimble 32, having a permeable bottom [240]. To perform a fast extraction, pipe 1 has in its right lower side a port that connects pipe 5 with pipe 4 by allowing a steam or hot solvent to enter/exit the big vessel by favoring extraction or separation and it is also a fusion point between the two parallel pipes (1 and 2). It establishes the communication with the pressure control valve A and with the component located at the lower part of the compact apparatus (see Fig. 6.2).

The experimental pressure of this compact extractor is manually controlled through valve A located at the right side of connector

4, which takes control on the upward hot fluid and/or the drain extract toward the reaction's flask or to the collector of sample in the receiver flask 31 (Fig. 6.2). Valve A also acts as a connecting bridge between pipes 1 and 2 as well as with the transporting pipe 5. Additionally, the transporting pipe 5 has a reduced cross-sectional area (optional) and is designed to form a preferred angle of 90° in the vertical direction to connect valves A and B but characterized to be short in its vertical portion to give strength and support for the other two parts. The parallel pipe 2 has an upper grade design (as shown in Fig. 6.2). On its upper side, section 11 is an upper joint available to accommodate a series of adapters or distillation components and be at the same time the specific point for an easy monitor of steam temperature that travels through pipe 10 or through the two segments of pipe 2 prior the fluid reaches the cooling chamber 9.

Thus, pipe 2 operates in such a way that steam (water vapor or organic gas) could reach the main extractor through the connector pipe 3 at observed high temperatures if steam or fluid travels down following the path of the enter/exit pipe 4. Then, the steam generated in the reaction flask could be directed toward extractor 1 in a straight line through the distillation pipe 10 to reach the pre-cooling chamber 23, and passing directly to the second condensing space 9 through connector's pipe 3. The segment 23 is a long and wide space designed to act as a temperature exchange chamber from where steam can follow toward vessel 1 through connector 3, or it is condensed to reach as hot solvent the sample like through pipe 4 (Fig. 6.3).

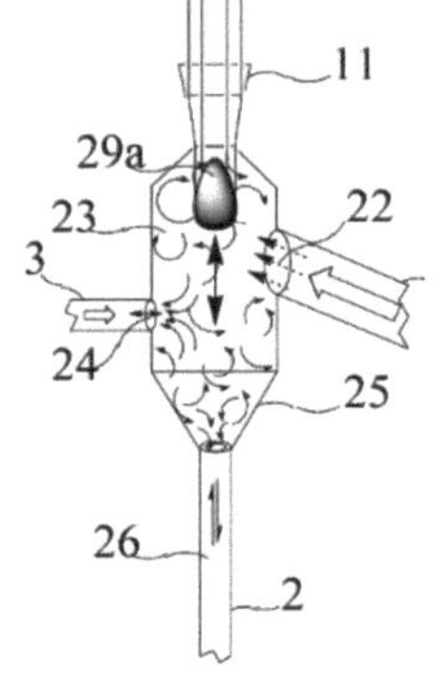

Figure 6.3 Scheme of chamber 23 where steam reaches pipe 1 through connector 3 or condense to percolate the sample in a countercurrent direction. Adapted from Ref. [238].

The temperature exchange chamber 23 plays an important role in the operation of SmCE. As shown in Fig. 6.3, if the port of communication 22 is constructed in a parallel plane or below the location of port 24, the overflow of the condensate generated in chamber 23 will end up returning to the reaction flask (30) through cylinder 10 without reaching the main extractor 1 if valve A is on position IV and valve B on position I, according to Fig. 6.4.

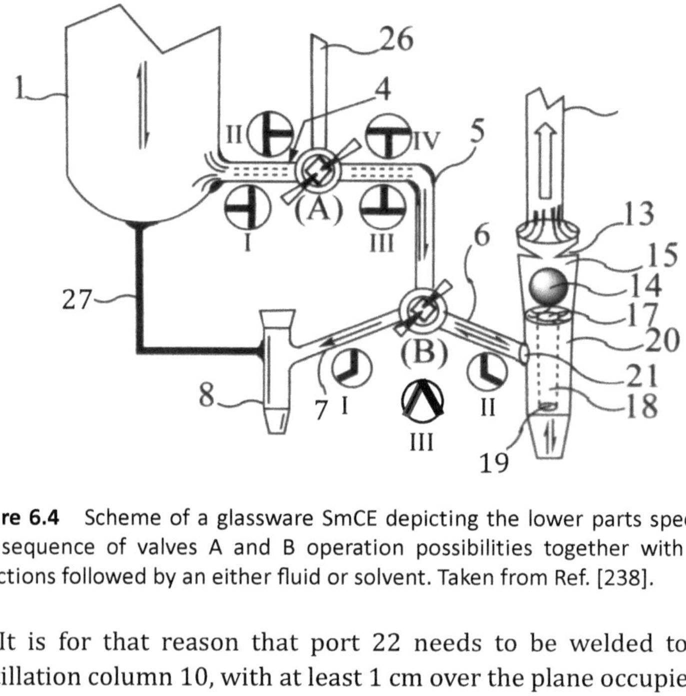

Figure 6.4 Scheme of a glassware SmCE depicting the lower parts specially, the sequence of valves A and B operation possibilities together with flow directions followed by an either fluid or solvent. Taken from Ref. [238].

It is for that reason that port 22 needs to be welded to the distillation column 10, with at least 1 cm over the plane occupied by port 24 in the opposite side on connector 3, as shown in Fig. 6.3, in order to balance the hydraulic pressure of the extract and the solvent, with the later having the possibility to descend through the conical and reduced sections 25 and 26, respectively, to reach the extractor through connectors 3 or 4, depending on the phase in which the fluid substance is in chamber 23 (a condensate or steam). Furthermore, design 23 deserves special attention because inside this segment, key physical phenomena can occur, such as:

1. Mass transport, as liquid and gas are formed in opposite ways, and
2. Heat transport, because liquid and gas gain or loss energy constantly.

These dynamic activities work in favor of a rapid phase transition, which occurs because the steam traveling upward through pipes 10 or 2 is subjected to quick change of pressure inside chamber 23 as a consequence of fast collisions between different quantities of steam molecules, which are flowing at different pressure and temperature by forming little cyclones and reaching chamber 23 through ports 22 and 25, with the atmospheric pressure entering through pipes 24 and/or 11, but being greater the pressure excerpted by the fluid entering upward through cylinder 10 to discharge in chamber 23.

Thus, the flow of a solvent with high density that reaches chamber 23 through the distilling cylinder 10 may condensate quickly and will overflow toward extractor 1 through the elongated segment of reduced diameter 26 (Fig. 6.2), which will be pushed through the semi-stationary mass of solvent in the conical section 25, plus the action of atmospheric pressure entering through port 24 and the pressure of steam raising inside column 10, with all these components excerpting maximum pressure over all areas in that chamber, as shown in Fig. 6.3.

With respect to the paths a fluid can follow inside a dynamic extractor, the Ch-G design allows for an inverse flow of the extractant fluid against the hydraulic pressure of the condensate that descends within vessel 1. Unlike as did not occur on any other similar glassware extractor, the fluid generated in the Ch-G design can flows in simultaneous counter-direction as an asset in the improvement of solvent extraction capacity for this extractor when limitations are related with a solvent that has high vapor pressure at low temperatures. In this regard, the extraction is carried out with a solvent that falls within the extractor vessel 1 as condensate, but also condensates quickly in the precooling chamber 23 generating large amount of liquid fluid that in the conical space 25 descend with high pressure to achieve the main vessel 1 through port 4 with this fluid being accelerated for an extra impulse due to hot steam is entering throughout the "T" stopcock A as counter-direction flow. Thus, the high pressure created in the reduced segment 26 causes that a great volume of solvent drips through the sample in opposite direction to the atmospheric pressure which is entering through condenser 28, by establishing a balance of phases between the densities of pure solvent and the solute extracted due to the high pressure that excerpts the mass descending throughout the narrow pipe 2 is by far greater than the constant mass of sample placed in chamber 1.

Therefore, the engineered design of these two parallel pipes for solvent fluid condensation and extractors' pressure balance, helps to solve the bottleneck presented by glassware extractors that requires a plurality of accessories that allows them to achieve the pressure balance between the fluid flow and the extracted fluid when the densities of the sample and the solvent are extremely different as usually happens [238]. In addition, on the right upper side of the same pipe 2, a connection port 22 establishes the first point of connection with the distillation column 10 in a preferred angle of 75°. This angle choice allows the steam to reach the pre-cooling chamber 23 at high pressure because previously the steam must have raised the mass of sphere 14 to continue its path through the distillation pipe 10. Thus, when a steam traveling through pipe 10 or the system of pipes on 2 reaches the chamber 23, it is a both high pressure and temperature, so then, peripheral accessories can be ensembled on port 11 as depicted in Fig. 6.2 (29a–c) regarding the process control or experimental determination of one of these two physical dimensions. The condensation space 9 can be occupied by a chiller accessory containing an arbitrary cooler that provides the necessary low temperature for the condensation of steam that that eventually will dropwise over the sample placed in a thimble or directly inside the extraction pipe 1 (Fig. 6.2).

Another advantage presented by the Ch-G design is related with the second pressure control valve B (made of Teflon or fritted glass), with its design to form a preferred angle of 120°. This geometry allows (1) one of the segments, connects B with the collecting exit pipe 7 towards its left side, (2) that B stablish connection with pipe 5 upwards, and (3) makes that B connects with chamber 20 through pipe 6 in its right side to give strong stability in the downward side and makes the Ch-G a compact and robust structure. A special part in this section of the Ch-G extractor is connector 7, which can be used in extraction or separation processes under continuous or step-by-step stages to collect aliquots of intermediate or final product for laboratory analysis through receiver 8, as shown in Fig. 6.1. Furthermore, connector 7 is a key accessory that can be used to introduce an inert gas into the main extractor or throughout the reaction flask when necessary or could also be the exit for solvent evaporation under conditions of medium to high vacuum (Fig. 6.5).

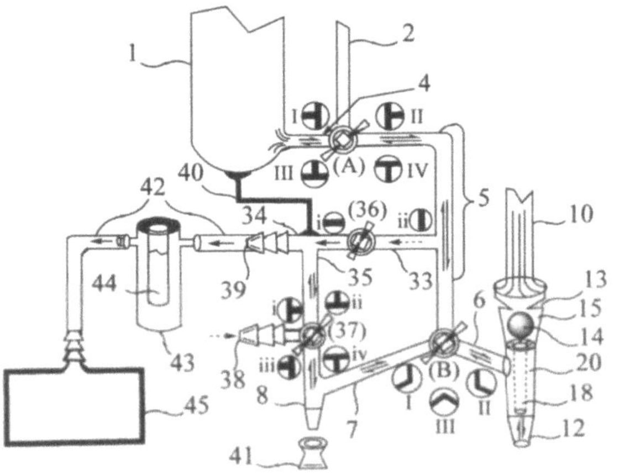

Figure 6.5 Schematic representation of additional accessories plausible to be welded on a Ch-G extractor for solvent evaporation or to feed a gas or apply medium to high vacuum in SmCE processes. Taken from Ref. [238].

In Fig. 6.5, the countercurrent connector 6 acts like the second point of fusion between the lower right side of pipe 2 and the left side of chamber 20, where the steam generated constantly in flask 30 can be guided toward the main extractor flowing initially through valves B (position II) and A (positions II, III, or IV), and entering extractor 1 by connector 3 if (a) the fluid reaches the upper side of pipe 2 at high temperature because the pressure of the steam flowing upward through the reduced diameter connectors 5, 26, 25 achieve the precooling chamber 23 with the steam deviated to the left side to condense on 28 and dropwise on the sample, or (b) the steam reach the extractor 1 as hot liquid fluid through connector 4 with temperature determinate by an either thermometer (29a in Fig. 6.2) or thermocouple (Fig. 6.1) placed individually on port 11 or as part of additional appliances necessary to carry on extraction with solvents that presenting high vapor pressure.

The Ch-G design permit that the extraction fluid achieves the upward part of the main chamber 1 throughout preferentially routes allowing that steam can be feed throughout the distillation cylinder 10 or throughout a conjunction of pieces within a set that follow the paths: 6 → B → 5 → A → 26 → 25, by doing that the fluid will be cooled on chamber 22 before reaches the extractor 1 through connector 3 to condense in chamber 9 and percolate the sample with large volume of solvent.

For instance, the selection of one of these two routes for solvent feed will cause that fluid at different rates enrich the main chamber 1 at different temperatures because feeding the fluid through the reduced diameter cylinders, it will be at low temperature when flowing upward so quickly achieve the extraction chamber. This route is a preferentially choice to balance the ratio of relation that appears between the low density of a light solvent and a high density of a sample due to the solvent's saturation in short time.

Also, the Ch-G extractor can be used for extractions carried out as a continuous or step-by-step process if one or more solvents with differences in boiling points of at least 10°C are placed on 30 and the steam generated is guided to enter extractor 1 by taking control of valve A. Thus, when this valve is opened downward (position IV), the solute dissolved in the solvent is removed by gravity by making sure that valve B is positioned toward its left side (position I). With this manual operation, the substance is drained toward connector 7 to be separated in flask 31 clamped in an adapter like 8, whereas a continuous extraction will be maintained if valve B is moved toward its right side (position II) with the extracted substance flowing continuously through pipe 6 by passing through space 20 downward the reaction flask 30.

In this case, the efficiency of an extraction must be favored if the solvent selected has a density greater than the sample and the accessory placed on the upside 11 port is only a thermometer (Fig. 6.2, 29a), while the process will be favored in a case where solvents with density lower than the sample are the choice and if the accessory on 11 is a Claisen adapter surmounted by 29b or 29c pieces, as shown in Fig. 6.2.

By using a Claisen adapter (see Fig. 6.2, 29c) with its complementary elements (a and b), the extraction can also proceed with the fluid or steam at mild temperatures, which will favor the extraction with solvents of lower density than the sample.

Figure 6.2 offers an illustration of what will happen in an extraction if valve A is in position I (opened toward extractor 1) and valve B is also in position I (closed for the stem of chamber 20).

In these conditions, a great volume of steam is quickly directed toward the pre-cooling chamber 23 through cylinder 10, dripping first on the sample as condensate generated in chamber 9, followed by a countercurrent solvent that percolates upward the sample and the extract is concentrated until reaching a level determined by the operator because of the movement of the solvent (gas/liquid) within the parallel pipes in a very short time. Conversely, if both A and B valves are open toward their right side (position II), a large amount of extract may be concentrated in extractor 1, which will be removed when B is opened toward its left side (position I) and recovered in the collecting pipe 8.

For a continuous process, the Ch-G apparatus can be set to works in an indefinite period with the one batch procedure dependent on the volume of solvent placed in the reaction flask according with the time setup. Therefore, the Ch-G extractor and the positions of both A and B valves according to the compact engineering design confer to any extraction or separation processes a special advantage due to the easy access an operator has over parameters such as:

- Solvent fluid direction.
- The system's pressure.
- Collection/drain of extract.

A key device in the SmCE construction corresponds to a glass sphere 14 (preferably colored) whose form, mass, and position introduce unique engineering particularities to this design, such as the possibility of directing the feeding fluid (steam water or an organic gas) by following a predetermined direction to balance the density of the extracted substance and the density of a solvent throughout a simple and safe way, because the design creates a seal point between surface 16 and the sphere in the upper ward port 17.

A key function of this physic device is directly related with the solvent path and direction which allows to it can be feeded (a) through the reduced connectors on 2 to reach the extractor 1 through either pipe 3 or 4, that occurs at low temperature or (b) as steam that travel through pipe 18 which raises into pipe 10 to reach the chamber 23 at very high temperature and pressure before condensing on 28. If valve B is closed (position I, Fig. 6.4), the hot steam reaches the condensing chamber 20 through connector pipe 3 and flows from the bottom of 18. Therefore, the sphere is lifted by the steam and its elevation will depend on the speed and pressure of the fluid created in 30.

The force of agitation observed in the sphere indicates, in simple words, the volume of steam flowing within the two chambers (18 and 20), which will allow an operator to take quick control of the experiment by regulating (a) the flow's direction of an either hot or warm steam by manually operating the two valves (A, B), and (b) taking control also on the pressure and direction of the steam even before the solvent has initiated its heating in the reaction flask.

Thus, the introduction of this physical component (sphere) into the SmCE apparatus addressed limited operation possibilities encountered with similar glassware extractors as the device operates over the pressure of a steam or fluid through a thermodynamic inequality expressed as:

$$Sphere_{\text{mass}} + P_{\text{atm}} + fluid_{\text{mass}} + g^* >>> P_{\text{vap}}$$
$$\text{whiting } [6, 5, 4, 26, 25, 24] \text{ sets} \tag{6.6}$$

**Gravitational force* [241]

Therefore, when the steam of a solvent occupies the conducting chamber 20, the lateral port 21 is available to deviate the steam generated at low temperatures toward the set of cylinders 6 → 5 → 4 → 26, reaching extractor 1 at low pressures, or to be feed upward through the distillation column 10 at high pressure and temperature when valve B is closed (position I).

Figure 6.4 shows an illustration of the coordinated operation of valves A and B with the vertical movement of sphere 14 within chamber 15 because of a positive pressure of steam that flows upward through the reduced section in cylinder 18 and whose temperature will be very high when the steam strikes the interior surface of the sphere.

If during an experiment, valve B is in position I (Fig. 6.4), the pressure corresponding to the steam raised by cylinder 18 will be sufficiently strong to lift the sphere until its limits, defined by a set of sharp designs (intruders), which forbidden the sphere to flight with the steam to a long distance inside chamber 15 and eventually fall on the concave surface 16 with negative consequences.

As described in the inequality of Eq. 6.6, the fixed displacement occurring on the sphere is due to the change in pressures between the two adjacent chambers by the pressure exerted upward by the hot steam in constant formation within chamber 20, which should counteract against:

- The weight of sphere 14,
- The atmospheric pressure in pipe 15,
- The mass of fluid condensing on the spere 14, and
- The gravity force acting on 14

Therefore, when valve B is closed (position I), it is thought that very hot fluid will surround sphere 14 by forming a turbulent flow that will make it turn in the fluid, which will initially condense on the surface of 14 helped by the presence of the intruders constructed on the inner side of the chamber. Therefore, fluid molecules with greater temperature will end up raising through column 10 by reaching chamber 23 and the extractor vessel 1.

Contrary to that, a change in the position of valve B toward its right side (position II) and moving valve A to one of its positions II, III, or IV will cause a drastic fall in the pressure of steam inside chamber 20. As a result, the pressure in chamber 15 becomes very high, which immediately deviates the direction of the steam toward the set of reduced section of cylinders 6, 5, 4, 26, 25, and 3, respectively. In addition to the above description, the Ch-G apparatus includes another key design related to one angular pipe engineered to be outside the limits of fusion between the two main parallel cylinders 1 and 2; and it refers to the *distillation cylinder 10*, that communicates the pre-cooling chamber 23 with vessel 1 through port 22 in its upper side and is fused in its below limits throughout connector 6 by doing of this part a third parallel cylinder. This embedded construction confers to the Ch-G design engineering capabilities (e.g., flow path, flow pressure, flow temperature) and compact characteristics.

This design has two points of connection; one through port 22 in its upper side that fuses the distillation tube with the vertical segment 23 on the pre-cooling pipe 2, whereas its lower side is fused with the upper side of chamber 15 over the intruder 13 by forming a compact extraction piece.

Moreover, port 22, as shown in Fig. 6.3, is a design forming a preferred angle of 75°, which acts as the entrance port of the hottest steam.

Figure 6.5 shows the front view of the farthest portion of a modified apparatus comprising the lower side of cylinder 10, the reaction flask 30, and the connection port 21. This additional arrangement (if necessary) is a complementary design of the Ch-G extractor possessing multiple manual parts to take advantage of the control of flow through the valves A and B and its direct action on the displacement(s) of sphere 14, because of the different pressures occurring between the chambers 15 and 20; and the existing pressures inside the pipes with greater diameter like 10, 23, and 1, respectively.

This last section of design is a compact piece, which consists of a great chamber space (15), which is flanked above by the end of pipe 10. It has three intruders to maintain the sphere inside the chamber, very similar to the intruders that are found in a Vigreux column, whose function in the Ch-G extractor is to prevent the glass sphere from moving away from base 16, which has a concave base and is, in turn, the lower limit of the chamber, opened to an entrance port 17, which connects with the adjacent chamber 20 through an internal pipe 18, and opens in its lower end 19 in this chamber.

The lower end of the piece is formed by chamber 20, which communicates with the reaction flask 30 through its polished conical end 12. In the left wall of the chamber, an entrance/exit port 21 is found, which is the final seal of the extractor components, where it merges with the pressure control and flow valve B through the connector pipe 6.

The current Ch-G design allows the use of an inert gas, which could be introduced through the collector device 8 when valve B is placed in position III, as shown in Fig. 6.4, and valve A is directed toward its left side (position I), with the gas entering extractor 1 through connector 3 with space 9 acting like a pressure-regulating area.

Furthermore, if both valves A and B are constructed in a boron-silicate material (plug and leg) and the upper sides of cylinders 1 and 2 are hermetically sealed, the extraction of a substance could be performed under partial or high vacuum by using the output device 7 in Fig. 6.5 with the turning piece in both valves directed toward their left side (position I) as shown in Fig. 6.4, but this time connection 8 is attached to the vacuum pump 45.

The manipulation of the two pressure control valves A and B allows the operator to remove the large condenser 28 from extractor 1 at any time without danger, when a process requires continuous cleaning of the apparatus for periodic replacement of fresh samples, except when it is operating under vacuum conditions.

Thus, successive operations of the Ch-G apparatus do not require a re-regulation of the condenser by allowing the extractor vessel (1) or the thimble part (32) to be filled with a fresh sample at any time during an operation without the need to disassemble the apparatus.

6.5 A Special Design for Moderated-to-High-Pressure or for a Vacuum Process

Figure 6.5 shows an additional design that makes a Ch-G extractor a tool for normal to ultra-vacuum operations necessary when volatile substances or toxic gases are generated, and their recovery is possible in an external storage system (like a pressurized tank).

An exemplary description of how an operator can manipulate the set of valves in an efficient and safe way is as follows:

For moderate pressure or vacuum:

- The sample must be placed directly in the extraction cylinder 1 or in the reaction flask 30.
- The outer joint of 9 and 11 to accommodate 28 and 29c cooler adaptations must be sealed.
- By using port 38, a weak-to-moderate flow of inert gas (e.g., Ar, N_2) will be introduced with valve 37 in position IV, and connector 8 must be sealed in its downward port (part 41) and valve B open downward (position III).
- Furthermore, valve A should be in position IV during the process.
- To complete a vacuum operation, valve 36 would be fixed in its horizontal position (I) by allowing the material of interest to be evacuated through segments 33 and 34, and to finally exit through port 39 by the action of a vacuum pump to be cooled in an external trap (parts 43, 44).

For high vacuum:

- The same arrangement and procedures as described in steps 1 and 2 of part A.
- Valve B may be fixed in its position III, and valve A in position IV, with the terminal of pipe 8 sealed too.
- The ports 38 and 39 can be connected to a high-vacuum (45) system with valves 36 on position I and 37 on position II, respectively.

6.6 Precautions while Using Glassware for Vacuum Procedures

6.6.1 Personal Protective Equipment

Personal protective equipment should be used when working with systems under vacuum conditions. Thus, for protection from potential flying glass debris, safety glasses, goggles, a Plexiglas face-shield, or a blast shield must be used when working near an apparatus under vacuum for long periods.

6.6.2 Tubing and Glass Equipment Assessment

Always that are used tube shape containers (e.g., metal, thick-walled rubber) that are rated according with their pressure or vacuum limit and special conditions you may follow their vendor warnings. For example, you must be sure that the tubing under use is in good condition and free of cracks, holes, spots exposing corrosion, or other signs of degradation. Never use glassware that is thin and scratched as well as shows clear star cracks. Be careful with the little defects observed on glassware, which are the weak points (little deflections) where it can easily break. Avoid as much as possible the use of round bottom flasks over 1.0 L in volume or glassware/equipment that is not approved for vacuum work [242].

6.6.3 Proper Checking of Connections

All connections (e.g., a ball joint, ground glass) in a system working under vacuum conditions require the use of clamps (e.g., keck, hose, pinch) to ensure the system integrity and safety. Before vacuum is applied, be sure that all ground glass connections are properly greased and glassware stopcocks are lubricated and properly sealed. Do not forget that any leak in the system can cause the vacuum source to break or cause the oxygen to condensate inside the system, leading to the plugging of narrow pipes if the system is cooled by liquid nitrogen [242].

In addition, a substance can be quickly dried by the action of a moderate vacuum or under an inert gas flow if it is contained in a one-neck flask, placed in the lowest outer joint of collecting tube 8 and stopcock B directed toward its right side (position II, closed).

A fast dry process could be achieved whether an additional vacuum system such as that observed in Figure. 6.5 is attached to the regular Ch-G design in its below left side with valve B in position II (closed), valve 37 in position **iv** connected to a vacuum pump 45 through port 39, respectively.

In addition, if a chiller or a cold finger (part 44) located inside a trap (43) can be coupled to segment 34 via connector 42, a vacuum pump (45) is also coupled to the cooler/trap arrangement via another connector 42 to complete a moderate to high-vacuum system.

Thus, the arrangement of Fig. 6.5 is an optimal vacuum toll for extraction or separation of light substances contained in biological samples such as aromas in flowers, light organic molecules as well volatile polluting agents dissolved in water.

Finally, two security glass rods (parts 27 and 40) contribute to secure the farthest portion of the apparatus comprising (a) the reduced pipes 5, 6, 7; (b) the control stopcocks set of A, B, 36, 37; and (c) additional ornamentation for vacuum operation, with the lower part of the main extractor 1.

As stated at the beginning of this chapter, the Ch-G apparatus is recommended to be constructed in a preferred embodiment of glass to handling biological sources (e.g., tissues) by doing of glass a more satisfactory and safety material for all the parts that has contact with the extract, except for the stopcocks A and B that could be made in Teflon for a biological (vegetable or organic) sample, or as borosilicate glass shell, with a plug which effects a seal (for vacuum) and ensure the purposes that characterize the semi-critical extraction method.

Thus, the compact extractor to carry out SmCE experiments is characterized to have a main pipe 1; a pre-cooling parallel pipe 2; a distillation pipe 10; the feeding lines 3, 4, 5, and 6; the reaction flask 30; the glass sphere 14; the collecting pipe 8 and its connector 7; a receiving flask 31; a stopper 41 and all parts made in glass that facilitate the cleaning of the equipment as well as preventing any chemical activity that can result if using other materials such as metals.

Nonetheless, many varying and differing embodiments can be made within the scope of the innovation conceptualized and described herein; and because advantageous modifications can favor the embodiment detailed here in accordance with the descriptive requirements of the law, it is to be understood that the details expressed herein are to be interpreted as illustrative in the benefit of the extraction process and not in a limiting sense.

6.7 Thermochemical Extraction of Biological Molecules through Semi-Critical Assisted Solvent Method

The extraction of folk medicines, lipids, fragrances, fatty acids, net proteins, vitamins, antioxidants, and metal ions with their natural properties is quite challenging in food, phytochemistry, medicinal chemistry, and biochemistry.

To date, most extraction methods that apply for food and medicines are:

- Protein cell disruption,
- Extraction/fractionation,
- Mechanical homogenization,
- Osmotic,
- Chemical lysis,
- Solubilization/precipitation, and
- Centrifugation.

A large number of them, however, fail to attain reliable amounts of entire molecules possessing their net biologically structure and function. Using the Ch-G compact extractor, we have developed easy extraction/separation of net proteins, lipids, enzymes, vitamins, and sugars from a broad scope of sources such as egg-white ovalbumin/yolk; cow and pork (beef), fish, cereal grains, and milk. In other experiments, antioxidants, fragrances, and lipids from vegetables sources have been extracted through the Ch-G upgraded apparatus, which improved yield and quality of chemical components operated at mild to semi-critical conditions.

At difference of other methods that requires the presence of a co-solvent or are initiated by co-precipitation, the design of the Ch-G allows to carry out an extraction under semi-critical conditions by using two solvents that has different physical properties (e.g., density, pH, polarity, vapor pressure) each one to be used together on the same batch for the extraction of molecules structurally dependent of temperature. The Ch-G engineering operation allows for an easy control of this parameter by its measurement in situ but that is difficult to detect with other glassware extractors.

Thus, the upgrade engineered parts introduced in the Ch-G apparatus, not only improves extraction in a continuous percolation which can be setup to occur in opposite pathways, but also give a way for the fluid's temperature or pressure be manually controlled as well as measured in situ by an either digital or analogic instrument placed on port 11 due to the special design 14 works synchronically through the manipulation of the two stopcocks (A and B) as an advantage on data collection of these two thermodynamic parameters that dominate any extraction process [238].

The **Ch-G** apparatus as shown in Figs. 6.1 and 6.2 can be constructed at different scale that includes pilot projects (academia, industry) to scalable size (high technology) by taken into consideration the type of sample or product of interest. The extractor allows for an easy extraction of net structure of proteins, amino acids, vitamins, chiral molecules, natural dyes, light carbohydrates, which biochemical and chemical integrity can be determinate through analytical methods such as HPLC, GC-MS, UV-vis, FTIR, ^{1}H-, ^{13}C-NMR, Raman, Fluorescence, Circular Dichroism, etc. [243, 244].

Furthermore, less time consume, and solvent amount are dramatically reduced (depending on the apparatus scale) due to the compact architectural design, which allows for a better and unique extraction performance if compared with the use of conventional (Soxhlet, Gregar) or expensive (Soxtec, Büchi speed extractor) technologies (see Fig. 6.6).

Figure 6.6 SmCE design for extraction/separation of casein, whey, and lactose from Bovine whole milk (2%) as an assisted solvent method on Ch-G extractor using citric acid (100 mL) at 40–45°C. Picture from author's lab.

6.7.1 Case Study One: Extraction of Two Proteins (Casein, Whey) and an Organic Sugar (Lactose) from Bovine Whole Milk 2%

Milk proteins are the nitrogen content of milk, which includes caseins (75%), whey proteins (18%), miscellaneous proteins (2%), and non-protein nitrogen (NPN) (5%). The total nitrogen distribution in milk is commonly determined by the Rowland fractionation [245] method, which has two main steps:

1. Caseins precipitation at pH = 4.6: It separates caseins (heavy molecules) from whey nitrogen (soluble light).
2. Precipitation with sodium acetate and acetic acid (pH = 5.0). It separates total proteins from whey NPN ones.

The milk proteins precipitate because the isoelectric point of milk is changed during the addition of an acid, which creates repulsions between different types of dissolved proteins (e.g., calcium caseinate) and other organic molecules when the milk solution achieves a pH of 4.6. At this point, casein becomes insoluble in solutions with pH less than 4.6 because the physiological pH of whole milk is near 6.6, and so casein acquires negative charge at a pH of 4.6 and can easily precipitate as a salt. Therefore, the addition of an acid to whole milk acts on the negative charges on the outer surface of the micelle, which are neutralized as the phosphate groups are protonated, and makes the neutral proteins precipitate [246].

Experiment

In a 1.0-liter graduate cylinder were measured 230 mL of Bovine whole 2% milk purchased from a local market and used without further treatment. The milk was poured directly into the Ch-G extractor fitted with an octagon magnetic stirring bar with integral pivot ring (10–12 cm length) and placed over a stirring plate, as shown in Fig. 6.6.

A 200 mL round-bottomed flask is fitted with boiling ships, and 100 mL of citric acid (pH = 2.4) (natural lemon juice) is poured in it. The flask is placed on a heating mantle, sand- or oil-bath with the flask attached to the lower side of the extractor (port 12), as seen in Fig. 6.6. After 45 min under moderate stirring and controlled condensation of the warm citric acid, caseins are precipitated (as shown in Fig. 6.7A) and recovered in a beaker (Fig. 6.7B).

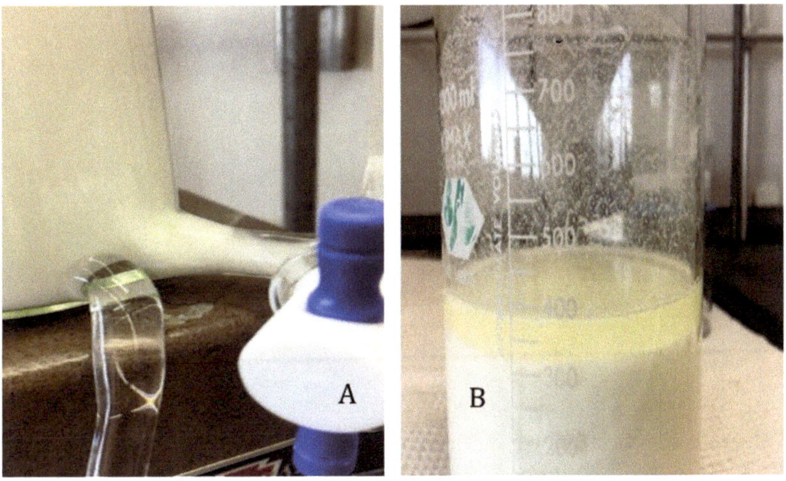

Figure 6.7 Casein precipitates after 45 min under the continuous addition of citric acid (A) at controlled temperature (35–40°C), followed by total precipitation and phase separation on a beaker (B) with whey proteins and lactose (upper layer solution). Picture from author's lab.

For research, the beaker content can be filtered by using a Thermo Scientific Nalgene Rapid-Flow Sterile Single Use Vacuum Filter Unit, or through a simple Hirsch funnel fitted with filter paper, connected to an Erlenmeyer flask with lateral adapter for water line vacuum filtration (see Fig. 6.8).

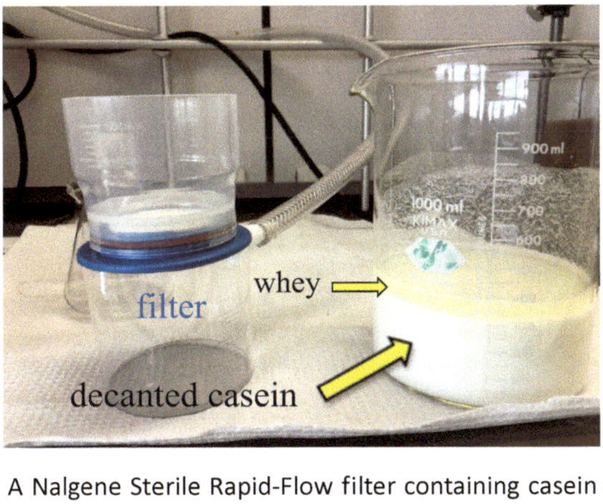

Figure 6.8 A Nalgene Sterile Rapid-Flow filter containing casein (upper side) and filtrate rich in whey and lactose (receiver flask 200 mL). Picture from author's lab.

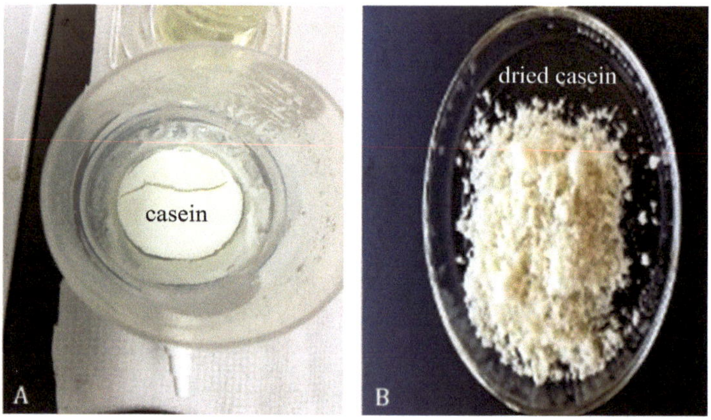

Figure 6.9 Bovine whole milk (2%) casein on filter (A) and dried (B). Pictures from author's lab.

After filtration, casein is retained in the filter system, dried thoroughly by vacuum or pressure, and finally placed in a dissecator to avoid humidity or high temperature (see Fig. 6.9A,B), which can interfere with the product's quality.

The recovered filtrate contains whey proteins and lactose for further separation. To do that, during the addition of an alcohol (ethanol or isopropyl alcohol — *never methanol*!), a cloud appears immediately in the mix with an accelerated precipitation of lactose, which can be improved if the recipient containing the mix is immersed in an ice bath for a few minutes before the alcohol addition (see Fig. 6.10).

Figure 6.10 Test separation of whey protein (A-I) and lactose (A-II, III) after addition of ethanol (99%) on the mixture and large lactose recovery (B). Pictures from author's lab. Data unpublished.

After lactose is fully decanted, it is separated from whey proteins, as done previously for casein, and dried. Microscopy can be used to determine size, shape, and appearance, as seen in Fig. 6.11.

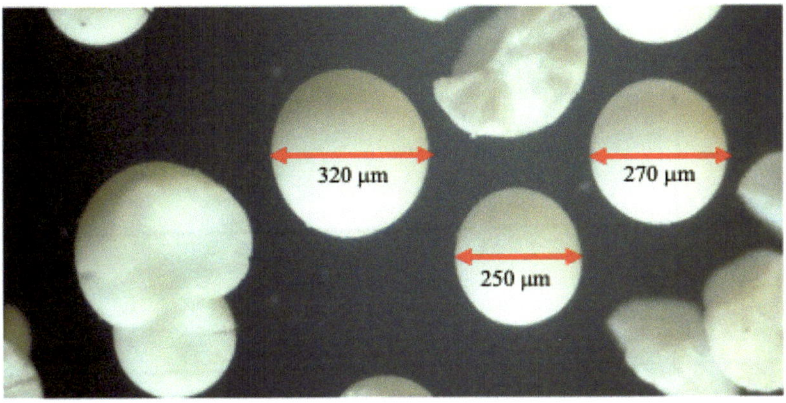

Figure 6.11 Micrography showing quasi-spheres of lactose as predominant shape of the sugar obtained in this case study by using bovine whole milk and citric acid as precipitant of proteins and sugar. Picture from author's lab. Unpublished data.

6.8 Extracted Proteins Brief Analysis

The yielded products for casein, whey, and lactose agreed with the data published by employing inorganic substances such as sulfuric or hydrochloric acids as isoelectric point changers that lead the net charge of these proteins to zero and precipitate [8]. The purity of products can be determined through spectroscopy UV–visible (UV–vis), FTIR, HPLC-MS, NMR as well as by circular dichroism, to analyze 2D and 3D structure of proteins. The infrared spectrum of the net sample was recorded on a Thermo Nicolet iS50 FTIR equipped with an ATR app. The UV–vis spectra were recorded on a Perkin Elmer Lambda 850 UV–vis spectrophotometer with sample dissolved in a buffer, as shown in Figs. 6.13–6.15. The circular dichroism spectrum was carried out from 190 to 350 nm and recorded in a JASCO 1200 spectrophotopolarimeter using a 0.01 cm path length cell at room temperature.

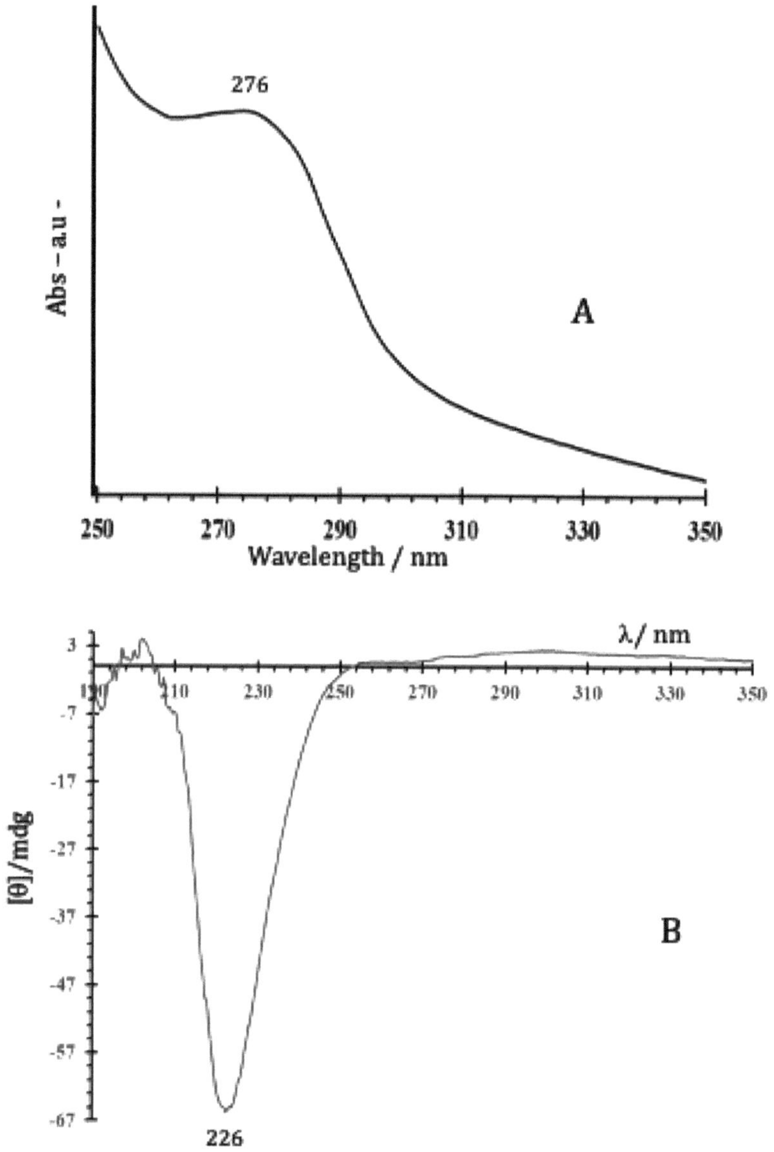

Figure 6.12 UV–vis (A) and CD (B) spectrum of *β*-casein from bovine whole 2% milk in sodium phosphate 50 mM, pH 7.5, 20°C (unpublished data).

Particularly, circular dichroism (CD) is an excellent method for rapid evaluation of the secondary structure, folding, and binding properties of proteins. Briefly, circular dichroism is defined as the unequal absorption of left-handed and right-handed circularly polarized light by a molecule that describes its natural or mutated activity [10].

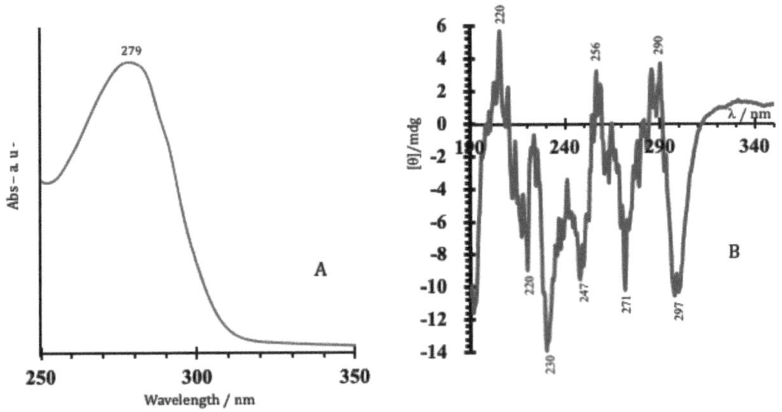

Figure 6.13 UV-vis (A) and CD (B) spectrum of whey proteins in sodium phosphate 50 mM, pH 7.5 at 20°C using a 1.0 cm cuvette (A) and 0.01 cm (B). Unpublished data.

All methods described to analyze CD spectra assume that the spectrum of a protein can be represented by a linear combination of the spectra regarding its secondary structure of elements, plus a noise term, which includes the contribution of aromatic chromophores and prosthetic groups, and mathematically can be expressed by:

$$\theta_\lambda = \Sigma \varepsilon_i S_{\lambda i} + \text{noise} \tag{6.7}$$

where θ_λ is the CD of the protein as a function of wavelength, the amount ε_i is the fraction of each secondary structure i, and the value of $S_{\lambda i}$ is the ellipticity (positive/negative in the plane) at each wavelength of each ith secondary structural element. In constrained fits, the summation of all the fractional weights, ε_i, must equal the unit (=1) and all the fractional contributions must be greater than or equal to zero (≥0) [247].

The CD data (Fig. 6.12) of casein extracted through SmCE, shows that the casein obtained has 67% of *beta*-casein conformation (β-CN), in excellent agreement with data reported on His-67 and Pro-67 caseins by doing of CD the best method to quantify the relative proportion of net β-CN variants in Bovine whole milk after the separation of its whey proteins by either the isocratic point difference achieve with an acid or through the enzymatic precipitation of

Bovine milks' casein [248]. A similar analysis, but performed for whey proteins (Fig. 6.14) and lactose (Fig. 6.14), shows accurate results for both molecular main conformations, as depicted in their corresponding absorption (A) and spatial configuration (B). From the collected CD data, it shown that semi-critical extraction carried out on the Ch-G apparatus, allows to attain proteins which preserve its key natural structure as evidenced for the high ellipticity (θ) values that characterize these valuable nutrition sources. Interestingly, it must be highlighted that 94.8% of lactose (Fig. 6.14B) is obtained as β-lactose as proteins and other net molecules are extracted from the milk micelles without changes in the sample temperature nor on the volume of acid added along the extraction process.

In addition to the above analysis, protein concentrations in the recovered samples can be determined using the respective molar extinction coefficients of the protein at 280 nm from its UV–vis spectra [247].

Furthermore, electrophoresis, HPLC-mass spectrometry (HPLC-MS) will be necessary for full characterization among other nutrition analysis. Thus, the Ch-G extractor shows suitability for the extraction of whole milk proteins in contrary to other glassware technologies, which are not available for this type of work due to their design, architecture, and affordability.

6.9 Case Study Two: Liquid–Liquid Extraction of Vitamin E from Infant Milk Formula Powder

Vitamins are water-soluble fatty acids that play key roles in human cells regarding metabolic and protection activity. They act as antioxidants and protect the skin against sunlight and skin diseases. These important metabolites can be extracted in the Ch-G extractor through SmCE at mild physical conditions. Thus, vitamins such as A and E can be afforded from infant formula milk powder, by dissolving a selected amount (gr) in 50 mL of deionized water, placed directly in the extractor's body to be extracted with 80 mL ethanol (99%) as solvent, in a liquid–liquid procedure.

The use of column filtration (solid phase extraction, SPE) methods has been recommended, which are performed throughout the Action Method 2011.08 for vitamin B_{12} in infant formula and adult/pediatric nutritional formula methods [249]. However, these late methods start with the warnings:

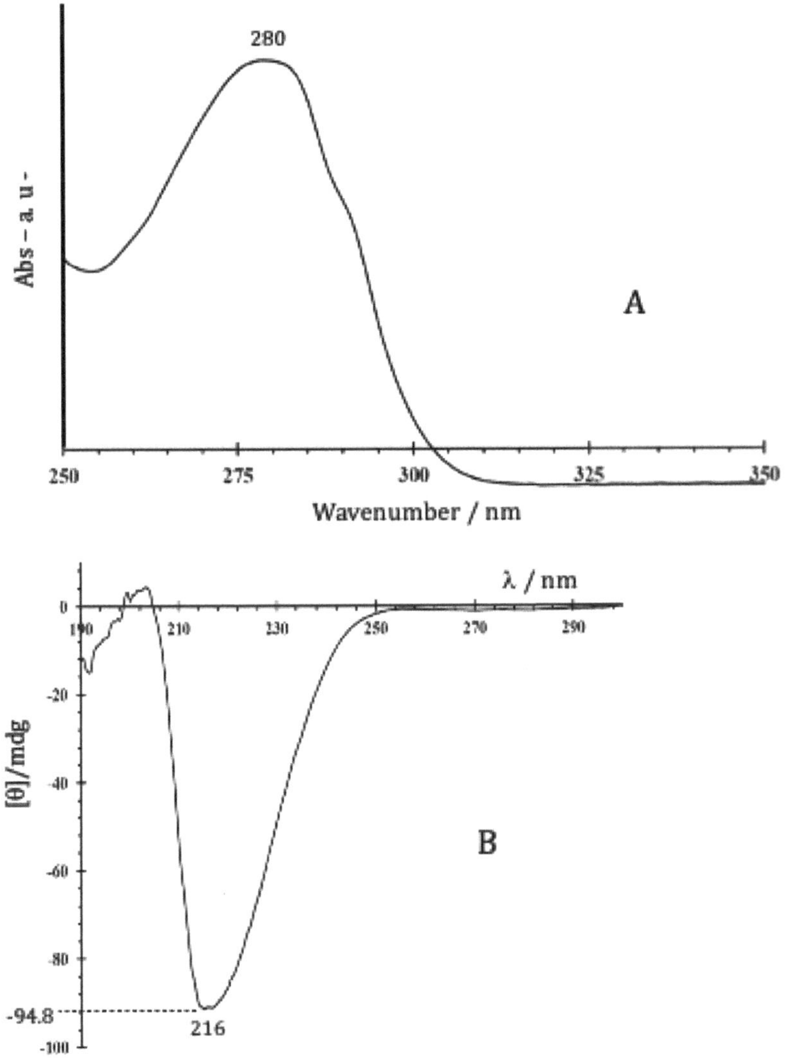

Figure 6.14 UV–vis (A) and CD (B) spectrum of β-lactose prepared in sodium phosphate 50 mM, pH 7.5 at 20°C in 1.0 cm path length (A) and 0.01 cm (B). Unpublished data.

Caution: Some of extraction methods uses commonly used solvents and reagents. Therefore, refer to appropriate manuals or safety data sheets to ensure that the safety guidelines are applied before using chemicals.

Cyanide — Fatal if swallowed, inhaled, or comes in contact with skin. Thus, wear protective gloves, clothing, and eyewear. Wash hands immediately after handling the product. Chemically, cyanide reacts with acids to form highly toxic and rapid acting HCN gas. Be sure you are using an effective fume hood removal device to remove all vapors generated. At the end of an experiment, destroy all residues with alkaline NaOCl solution.

Trifluoroacetic acid (TFA) — This acid causes severe burns and eye damage. Thus, wear protective gloves, clothing, eyewear, and face protection. Use only an effective fume hood removal device to remove all the vapors generated during the use of TFA.

The liquid–liquid extraction (LLE) of vitamins performed in the Ch-G extractor uses the LLE protocols, which take less time to complete the extraction in comparison to the spent time when performed through the SPE method in addition to avoiding the employment of hazardous reagents such as HCN and TFA. Although this last method is now experiencing a renaissance, the use of highly toxic volatile compounds makes this procedure more experimental and not suitable for clinical and routine use for high throughput analysis.

Figure 6.15 Scheme of liquid–liquid extraction of vitamin E in Ch-G extractor from infant milk formula powder in the Ch-G extractor. Unpublished data.

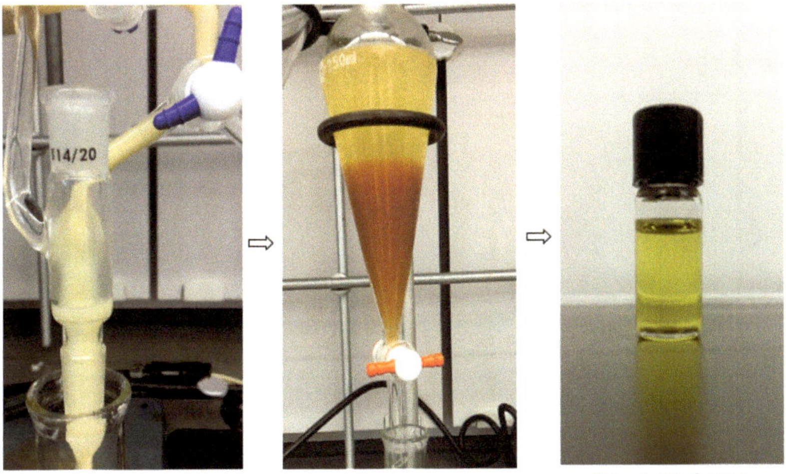

Figure 6.16 Scheme of product recovery, partitioning separation, and final product isolation (vitamin E). Unpublished data.

6.10 UV-vis Spectroscopic Analysis

Ultraviolet-visible (UV-vis) spectroscopy can detect the electron transition of the chromophore of pure vitamin E with an absorbance peak occurring at 290 ± 1.3 nm in nonpolar solvents (hexane like). The vitamin E extracted by LLE in the Ch-G extractor was prepared in hexane, and its UV-vis spectrum shows a set of bands in the ultraviolet region (UV), which shows four maximum peaks with height absorptions occurring at 270, 280, 302, and 320 ± 1.3 nm wavelengths and other significant trends in the near-visible region with respect to radiation absorbed by isomer vitamins in the sample [250].

This analysis thereof cannot be understood as conclusive, because FTIR, GC-, LC-, and HPLC-MS spectrometry are by far the AOAC and FDA methods universally accepted to validate the analysis of vitamins extracted through any preparation method [250].

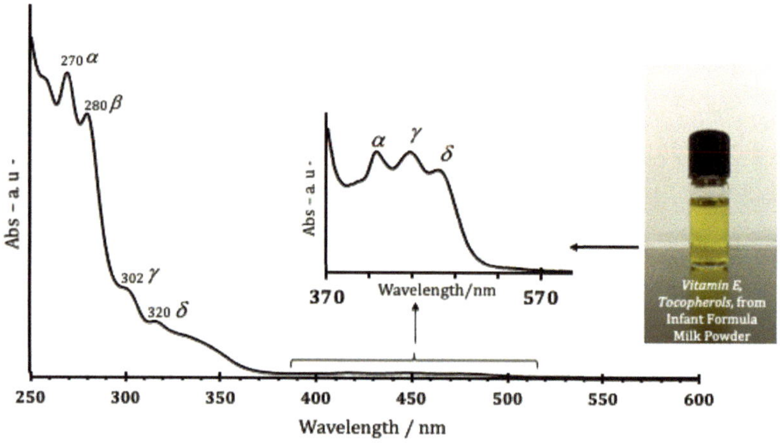

Figure 6.17 UV-vis spectrum of vitamin E extracted from infant formula milk powder through SmCE at 35–40°C. Unpublished data.

6.11 Case Study Three: Semi-Critical Assisted Solvent Extraction of Algae Oil

Algae are primary producers in ecological systems because they are widely distributed in salted and fresh water sources and are closely related with human ecosystem activities [252].

The biofuel synthesized from these photosynthesis species are often considered the *fuels of future* because they present an easy access to its farm production during a season-round cultivation, they have low-competition for land used if compared with food crops, has high calorific energy content, acts as ecosystems for environmental remediation. In addition, and at difference of fossil fuels, algae possess great potential to produce fuels directly under genetically engineered improvements [252].

But the access to algae biofuel using both conventional technology and chemical procedures has been costlier than that encountered for oilseeds due to the difficulties presented by these microorganisms regarding accessibility to their cell wall matrices [253].

Nonetheless, algae lipid extraction, have been dominated for several technologies that includes but are not restricted to solvent-assisted (Soxhlet), ultrasonication (liquid waves frequency dependent), microwave-assisted extraction (heat-mechanical), subcritical, and supercritical fluid extraction (SC, CO_2 dependent) discussed in Chapters 4 and 5 [254–258], among to direct conversion of algae (micro) biomass into biofuel by synergic thermochemical methods [259].

6.11.1 Macro Algae Preparation

For this case study, wild green algae, *Spirogyra* sp., was harvested at different points of a river located in the northwestern region of Puerto Rico Island, US commonwealth territory, with coordinates N 18°4'12.6516," W 66°59'53.646," as well as in a pond located at Howard County, Maryland, during the spring seasons of 2012–13 and 2015–17. The samples were dried at room temperature for 24–36 h until constant weight was attained. The dried samples were grinded through a mortar and pestle, and the lower size particulates were passed through a 500 μm sieve to remove oversized particles and submitted to continuous extraction.

6.11.2 Conventional Extraction

This extraction was performed on a Soxhlet conventional extractor (55/50) following published methods [260] by taking into consideration, for analytical purposes, the yield of crude oil, the experimental time, and the sample/solvent relation or the algae mass–solvent (w/v) ratio.

To test cell wall activation, several solvent systems were selected for an initial evaluation; however, an azeotrope of hexane and acetone in a 2:1 ratio was finally chosen to be the best activation mix for algae oil extraction. Thus, a precise mass of grinded algae (10 g) was placed in a thimble (glassware with porcelain fritted base) with dimensions of 25 mm ID × 80 mm height, and the thimble was placed inside the extractor, which is surmounted with a condenser and the vessel is suspended above a round-bottomed 400 mL flask fitted with boiling ships and 250 mL of solvent.

The extraction was allowed to proceed for 4 h under reflux with subsequent removal of solvent in a rotary evaporator set at 35°C and 110 mbar. At least five replications (each season) for each extraction were performed with the results averaged. The mass of recovered crude oil is compared to the initial dried algal biomass/solvent ratio to determine percent recovery as calculated using the following equation:

$$\%EOM = \frac{EOM_{total(g)}}{M_{algae(g)}} \times 100 \quad (6.8)$$

where %EOM is the percentage of algae oil mass; EOM_{total} is the total oil matter expressed in grams, which is the sum of all extracted and identified organic compounds in the product; and M_{algae} is the mass of ground algae sample subjected to extraction.

In several studies, linear solvents (e.g., n-alkanes, halogenated hydrocarbons, and alcohols) with different chain substituents can form two distinct polar and nonpolar units, which confer the lipophilic character to the molecule and act as a co-activant on the algae cell wall activation. The chemical effects of mixed polar azeotropes, on the other hand, have demonstrated good effectiveness for algal oil extraction/recovery when the solvents used are mixtures of n-hexane/alcohol in different ratios [261]. For similar algal-oil extraction utilizing an azeotrope of two solvents, a good choice would be the mix of a nonpolar activant like hexane with another polar like acetone for the lipid extraction utilizing a macro-alga as the source. The mix of these two solvents possesses more carbon-hydrogen atoms than the mix n-hexane/ethanol, nonetheless the density of the two azeotropes is the same but the cell wall activation achieved with the latter mix, makes of this a best solution that works in favor of lipid extraction.

6.11.3 Algae Lipids SmCE Assisted by Solvent

When extraction of algae lipids was carried out in the Ch-G extractor under the same conditions and sample amount, yields of biocrude oil were significantly enhanced from dried wild green *Spirogyra* sp. treated with either plain n-hexane from 4.91 ± 0.41% (in Soxhlet) to 16.80 ± 0.5% (in SmCE; Ch-G) and when extraction was on plain acetone, the yields were 5.44 ± 0.26% (Soxhlet) and 18.31 ± 0.21% (SmCE; Ch-G), which suggests that macro algae biocrude extraction and recovery are undoubtedly upgraded (fivefold) under SmCE rather if extraction is carried out using the traditional siphoning (Soxhlet) method which had reported yields very similar to that of employing either subcritical or supercritical fluid extraction [262]. However, the best extracted/recovery percentages of biocrude from this wild green *Spirogyra* sp. macro algae are those extracted with a mix of solvent composed by a polar and nonpolar solvent [263,

264]. For this, an n-hexane/acetone polar mixture (2:1) showed to improve the extraction efficiency for wild green *Spirogyra* sp. macro algae by using either Soxhlet or Ch-G extractor with lipid extraction and recovery efficiency ranging from 7.26 ± 0.45% (using Soxhlet) to 34.82 ± 0.83% (on SmCE; Ch-G) (fivefold), respectively. The obtained biocrude yields were determinate after liquid column chromatography (LCC) was used to its clean with their amounts summarized in Table 6.1. The hydrocarbon and fatty acid methyl ester (FAME) profiling by ^{1}H-NMR (Fig. 6.20b) and GC-MS (Fig. 6.21) shows that not only lipids with high quality were recovered in a short time (low energy consumption), but also they contain fewer percentages of inorganic contaminants (e.g., S, heavy metals), although the algae samples were collected from sources of different origin (Maryland and Puerto Rico). Elemental analysis and calorific value were in good agreement, as shown in the data presented in Table 6.2.

Table 6.1 Yield of crude oil extracted with pure and solvent mixture [n-hexane (H), acetone (A), and azeotrope (H/A)] through Soxthlet and semi-critical assisted solvent extraction methods

	Extraction method, Yied (wt%)[a,b]		
Algae	**Solvent**	**Soxhlet[c]**	**SmCA-Sol[d]**
Wild green	Hexane	4.91(± 0.41)	16.80(± 0.52)
Spirogyra sp	Acetone	5.44 (± 0.26)	18.31(± 0.21)
	Azeotrope H/A(2:1)	7.26(± 0.45)	34.82(± 0.83)

[a]Standard deviation, [b]4h, [c]250 mL, [d]80mL

6.12 Analysis of Algae Lipid

6.12.1 FTIR

The dark oils obtained through the SmCE method were simply cleaned through liquid column chromatography (LCC) by using silica as the stationary phase and hexane as eluent. Thus, the infrared spectrum of crude oil was recorded by using a resolution of 0.09 cm^{-1} and 32 interferograms were recorded on a Thermo Nicolet iS50 FTIR spectrophotometer.

Table 6.2 Elemental analysis, inorganic matter, and calorific values of crude oils of *Spirogyra. Sp* macro algae

	Element (%)[a,b]						Inorganic matter (ppm)[c,d]				Calorific value
Sample	**C**	**H**	**N**	**O**	**S^b**	**S^c**	**Na**	**K**	**Ca**	**V**	**(MJ/kg)[h/2]**
Crude-oil	72.56	10.59	1.01	15.46	1.10	0.47	46.2	135.2	20.4	105.8	40.55
Residual											8.50
Pure-oil							30.0[d′]	100.0[e]	20.0[d′]	2000.0[f]	28.0[g]

[a]Galbraigth laboratories, Knoxville, TN., [b]by pyrolysis, [c][ASTM D-4294], [d]AAnalyst 100 FLAME Atomic Absorption Spectrometer – [ASTM D-5863 B] method; [d][ASTM D-7303] method, [e](Denver et al., 2008), [f](vale et al., 2008), [g](Chaiwong and Kiatsiriroat., 2015), [h]Adiabatic Parr 2901EB Oxygen Bomb Calorimeter [ASTM D-240 method]. Reprinted from Ref. [26].

The IR sample was prepared by dropping a chromatographed pure sample (neat crude) over a round cell window of NaCl (25 mm diameter × 4 mm thickness) to form a thinnest layer with a circular polytetrafluoroethylene (Teflon) spacer hole with thickness of 0.05–0.1 mm. Another NaCl window is placed over the sample to form a sandwich with the two windows secured in a Wilks IR sandwich cell sample holder.

The FTIR spectrum shows characteristic stretching at 3395 cm^{-1} ascribed to O–H vibration as found in commercial biodiesel (3391 cm^{-1}; Vinita et al., 2013). A strong mode at 2919 cm^{-1} is assigned to the asymmetric C-H stretching of CH_2, CH_3 groups of aliphatic chains, whereas the corresponding symmetric vibration for these groups is a sharp signal at 2838 cm^{-1}, respectively. The strong stretching at 1701 cm^{-1} is assigned to carbonyl groups (–C=O) that are present on either esters (R–C=O–O–R′, R–C=O–R′O) and carboxylic acids (–C=O–OH) in large number on this *Spirogyra. Sp* crude oil.

The moderate stretching at 1274 cm^{-1} corresponds to the C–O deformation, which indicates the presence of ester carbonyl group, although other groups also absorb in this region. The presence of carboxylic acid as the functional group in these extractions is confirmed by the medium intensity stretch in the region of 1451–1370 cm^{-1}, which can be assigned to the bending of the hydroxyl (O–H) group.

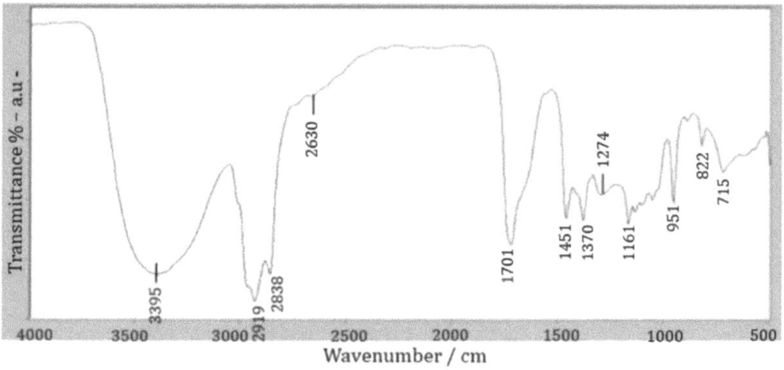

Figure 6.18 FTIR spectrum of net biocrude oil of *Spirogyra* sp. macro-algal prepared as thin film, 0.05–0.01 mm width between NaCl 5.0 mm windows. Reprinted from Ref. [263].

A weak mode at 1161 cm^{-1} is ascribed to C–O–C stretching, as found in other lipids [264]. Two weak absorptions at 1270 cm^{-1} and 822 cm^{-1} are assigned to C-H deformations. The stretching at 1370 cm^{-1} is assigned to C-H/O-H deformation, and the stretching with moderate intensity at 951 cm^{-1} is ascribed to $-CH_3$ rocking.

These two low energies stretching taken together demonstrate that both modes have chemical structures identical with those that characterize branched lipids, which have also aromatic/aliphatic rings attached to long hydrocarbon chains.

6.12.2 UV-vis Spectroscopy

On chromatographed algal oil, steady-state absorbance spectra were recorded on a Perkin Elmer Lambda 850 UV–visible (UV–vis) spectrophotometer with spectra collected on chromatographed oil dissolved in methanol, as shown in Fig. 6.20. A strong absorption at 380 nm is ascribed to all sample components with one of those contributing to the characteristic maximum of crude oil absorption in the visible range of spectra. Pure aliphatic acids, esters, and corresponding triglycerides are colorless substances; therefore, they do not exhibit significant absorption in the visible range, but natural fatty acids and oils from plants and animal pigments commonly exhibit visible absorption as has been found in the *Spirogyra* sp. sample [265, 266]. The presence of metal ions such as vanadium (V^{IV}) and copper (Cu^{II}) enzymes, which naturally occur in photosynthetic species, can be also detected in the visible region of the spectra below 500 nm, together with chloroplasts and other natural dyes [267, 268].

6.12.3 Lipid's ^{1}H-NMR Characterization

Proton NMR (^{1}H-NMR) is a sensitive analytical technique that allows determination of chemical purity and molecular structure of either organic or inorganic substances obtained by extraction or synthesis methods.

Thus, a chromatographed aliquot of algae oil was subjected to NMR analysis with its spectra recorded on a Bruker Avance III 400 MHz NMR spectrometer equipped with dual probe (^{1}H/^{13}C). The sample was prepared by dissolving approximately 5 to 10 mg of algae oil in 0.7 mL of deuterated dichloromethane (DDCM) CD_2Cl_2 containing the internal standard tetramethylsilane (TMS). To run an experiment, the instrument parameters such as relaxation delay (RD) and receiver gain (RG) were optimized, and 90° PW calibrated to sufficiently allow relaxation of nuclei and get the quantitative spectra. The structures of purified lipids can be confirmed by ^{1}H-NMR (Fig. 6.21), and the analysis of the results is summarized in Table 6.3. The chemical shifts in this case are assigned by comparison with former reported vegetable oils (Shimamoto et al., 2015), and these results must be in good accordance with structures checked by GC-MS, or HPLC-MS spectrometry techniques [269, 263].

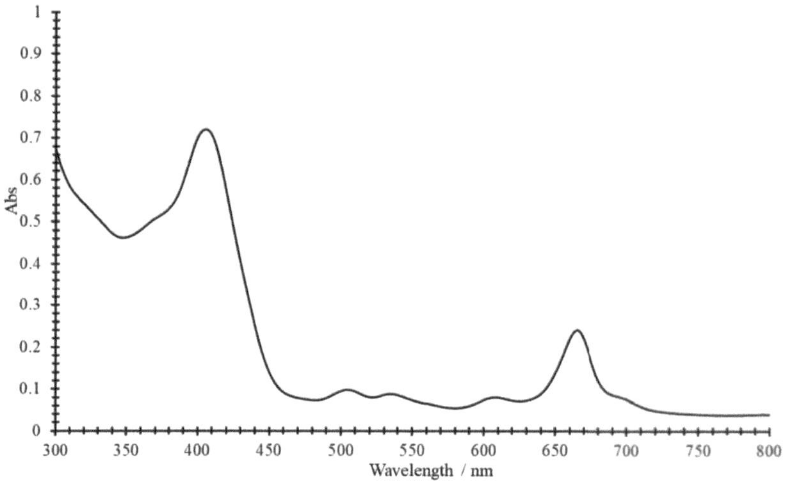

Figure 6.19 UV-vis spectra of chromatographed *Spirogyra* sp. crude oil. Reprinted with permission from Ref. [263].

Table 6.3 ^{1}H-NMR (400 MHz) chemical shift data for macro algae *Spirogyra* sp. crude oil

Chemical Shift (ppm)	Multiplicity	Assignment (^{1}H) (a-j)^a
0.889	Multiplet	-(CH$_2$)-**CH$_3$**; (j)
1.085–1.104	Multiplet	-CH=CH-CH$_2$-**CH$_3$**; (i)
1.224	Singlet	-**(CH$_2$)$_{3\text{-}5}$**-CH$_3$; (h)
1.581	Multiplet	-(C=O)-CH$_2$-**(CH$_2$)$_3$**-; (g)
2.021–2.114	Multiplet	-**CH$_2$**-CH=CH-; (f)
2.301–2.352	Multiplet	-(C=O)-**CH$_2$**-(CH$_2$)$_n$-; (e)
2.851	Multiplet	CH=CH-**CH$_2$**-CH=CH-; (d)

(*Continued*)

Table 6.3 (*Continued*)

Chemical Shift (ppm)	Multiplicity	Assignment (^{1}H) (a-j)[a]
4.140–4.180	Quartet	-C-O-**CH_2**-CH-; (c)
4.270–4.323	Two doublets	-C-O-**CH_2**-CH-; (c)
5.141–5.155	Multiplet	O-CH_2-**CH**-CH_2-O; (b)
5.261	Singlet	CH_2-(**CH=CH**)$_n$-CH_2; (a)

[a]as described in Fig. 6.20b. Reprinted from Ref. [263].

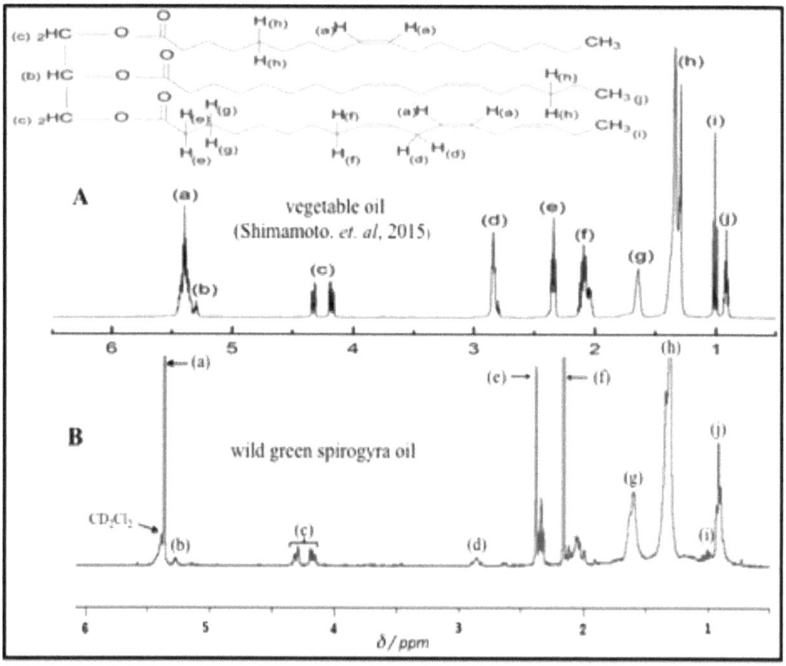

Figure 6.20 ^{1}H-NMR 400 MHz spectra of an exemplified structure of vegetable oil (A) with macro algae *Spirogyra* sp. crude oil (B) with assignment of hydrogen (by comparison). Reprinted from Ref. [263].

6.12.4 Gas Chromatography–Mass Spectrometry (GC-MS) Analysis

One of the more accepted techniques for chemical analysis of food, medicines, and phytochemicals is the coupled gas chromatography–

mass spectrometry (GC-MS) system, because of its accurate and reproducible results.

To evaluate the quality of lipids extracted from a macro alga through the SmCE technique, biocrude extracts were analyzed this time by following a method suggested in the literature [270]. Thus, the GC-MS measurements were recorded on samples of chromatographed crude oils and carried out on an Agilent 7890 GC system coupled to an Agilent 7000 GC/ MS Triple Quad spectrometer with an Elite-I fused silica capillary column (30 m × 0.25 mm ID × 0.25 μm (HP-5 ms)). The samples were prepared by dissolving 1.0 μL of algae biocrude in 10.0 mL of hexane (HPLC grade) with sample vortexed during 5 min and fractions filtered through 0.2 μm sterile syringe filters. For GC/MS detection, an electron ionization system with ionizing energy of 70 eV was used at scan intervals of 0.5 s and fragments from 45 to 450 Da with GC total running time as 30 min. The samples were used without derivatization (non-diluted samples recommended) and transferred as 200 μL of the prepared solution to the GC-MS sample vials as 5 μL of this solution and injected into a split/splitter injector with the oven setup at 250°C.

The oven temperature was set up as a ramp to change from 50°C to 150°C at a rate of 10°C/min (isothermal for 2.5 min), and after this initial data collection, a new ramp from 150°C to 270°C is setup to change at a rate of 5°C/min (isothermal for 30.0 min) to complete this experimental analysis.

Helium (He, 99.999% UHP) was used as the carrier gas at a constant flow rate of 2.0 mL/min, and an injection volume of 2 μL was employed at a split ratio of 10:1 and injector temperature 245°C; with ion-source isothermal at 270°C.

The relative amount (in %) of each chemical component is in situ calculated by comparing its average peak area to the total area, and the software adopted to handle these oil mass spectra data was a Turbomass package.

To evaluate the method (SmCE) efficiency and the extractor (Ch-G) efficacy in the extraction of lipids from algae cell matrices, the obtained oils were subjected to GC/MS analyses in order to find out hydrocarbons and fatty acid methyl esters (FAME) present in the crude oil of *Spirogyra* sp.

The data show that the oil is a mixture that contains predominantly groups of hydrocarbons and lipids, as shown in Fig. 6.21.

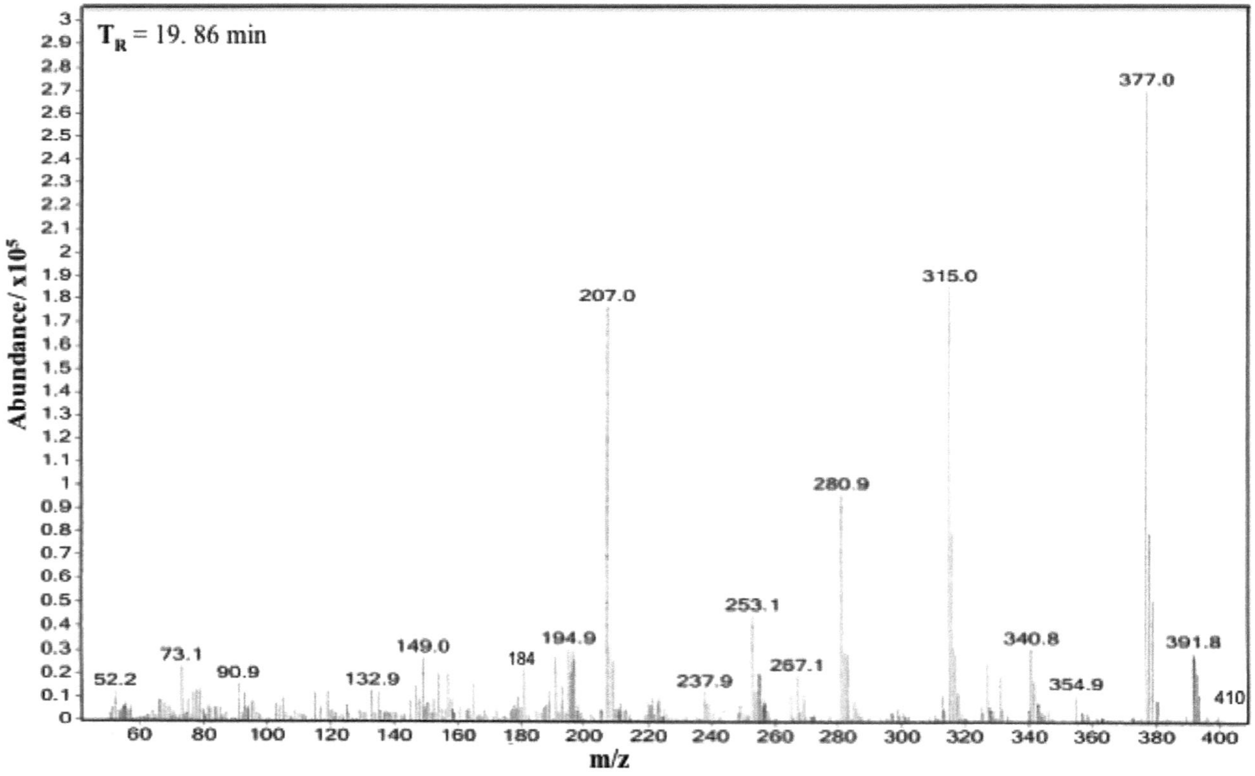

Figure 6.21 GC-MS spectrum of macro algae *Spirogyra* sp. azeotrope fraction with ion pattern for spectral peak at t_R = 19.86 min. Reprinted from Ref. [263].

The chemical structure assigned for the extracted lipids from macro algae *spirogyra. sp* are from a GC-MS data, which is based on the analysis of a molecule fragmentation pattern that is generated for the mass spectra by a direct comparison of the collected data with profiles of free access in the National Institute of Standards and Technology (NIST2) library, together with mass spectra of data published in the literature.

From the data in Fig. 6.21, it is easy to see five major peaks in the fraction oil obtained with azeotrope solvent at retention times between 12.64 min, 19.86 min, and 22.00 min. The peak at t_R = 19.86 min displayed a molecular ion at m/z 391.8, suggesting a structural formula of $C_{27}H_{51}O$. The loss of a methyl ($-CH_3$) group from the molecule resulted in a new m/z 377 mass signal that corresponds to a $C_{26}H_{49}O$ fragment.

Other characteristic ions include heavy ions such as one at 315 (M^+ $-C_{18}H_{33}O_4$+-2H*) typical of bis (2-ethylhexyl) ester in addition to a peak at 280 (M^+ $-C_{18}H_{31}O_2$) and other at 253 (M^+-$C_{16}H_{30}O_2$) associated with a sequentially fragmentation of saturated dicarboxylic acids, respectively.

The analysis of the peaks that occur at t_R = 12.64 min, 19.86 min, and 22.00 min revealed a fragment pattern ion at m/z 207 in that a heavy hydrocarbon with formulae $C_{15}H_{28}$ is eliminated, thereby yielding a (2Z, 6Z, 10Z)-3,7,11-trimethyl-2, 6, 10-dodecatrien ($C_{15}H_{27}$) ion.

A signal at m/z 166 is typical of an acyclic-type hydrocarbon with formula $C_{12}H_{22}$. The peak that appears at m/z 73 is assigned to be a light hydrocarbon (C_5H_{11}), which is an isobutyl ion. The ion peak at m/z 52 (C_4H_4) is assigned to be cyclobutadiene ion. These ions are characteristic of squalene fragmentation. The fragment with a signal at m/z 90 (M^+-$C_4H_9O_2$) corresponds to 2-methyl-propanoic ion. Other intense ions at m/z 194, 184, 149, 207 (base peak, $C_{15}H_{27}$), and 52 are characteristics of saturated and unsaturated alkanes.

6.13 Conclusion

From the several case studies discussed by either duplicate or triplicate data collection, the semi-critical assisted solvent (SmCE) method proved to be a reliable process to afford a high quality of hydrocarbons, proteins, antioxidants, and lipids extracted from a wide scope of sources, which included wild green macro algae, bovine whole milk, infant formula milk powder, to curve some paramount downstream processing difficulties encountered in the lipids, net proteins, and biology functional molecules production.

In addition to that, non-halogenated polar solvents showed best reliability for lipid extraction, as they are thought to provide the highest lipid recovery.

Although new proposed solvent-free methods appear to be promising at the laboratory scale, they need more research to be carried out at the commercial scale to minimize the use of hazardous halogenated solvents.

Our aim in this direction has been improvement in chemical extraction methods as reliable, less time consuming, and cost-effective production of high-quality proteins, amino acids, antioxidants, nutraceuticals (e.g., whey proteins, lactose), and liquid biofuels from macro algae within others.

The results presented in the case studies using aqueous and organic plain as well as azeotrope-mixed polar solvents in the thermochemical upgraded extractor (Ch-G) improved the yield of algal lipids when compared with extraction carried out in the traditional Soxhlet method.

This preliminary investigation found that plain hydrocarbon solvents are not enough as algae cell wall disrupters when compared with the effectiveness achieved with synthesized azeotropes possessing mixed polarity. Thus, the performance of SmCE technology assisted by solvents shows to be an excellent alternative for biofuel material synthesis and transformation, due to the improved chemistry and physical correlation between high percent of hydrocarbon extracted and calorific values detected without the algal lipid's purification, contrary as is used that lipid are analyzed only after esterification steps.

Therefore, the high purity of the hydrocarbon detected by proton NMR on the algae lipid gave a C/H ratio composition for $C_{15}H_{26}$ as main empirical formula for *spirogyra. sp* biocrude oil. The gas chromatography–mass spectrometry (GC-MS) analysis shows a chemical composition that includes unsaturated hydrocarbons and phenols, in addition to heterocycles containing nitrogen as well as long chain fatty acids and amides. In addition to the high percentages of hydrocarbons and oxygen, a low content of nitrogen, sulfur, and heavy metals was detected in the macro algae sample belonging to an urban pond as well a freshwater stream.

A detailed spectroscopic analysis of biocrude oils was undertaken using FTIR, which shows, in particular, the lipidic components containing oxygen in combination with nitrogen by forming aromatic nitrogen compounds and free fatty acids as predominant chemical structures.

The findings are not our concluded work but till address a significant improvement in algae cell wall disruption as a case study to obtain biofuel with high calorific content very close to or better than those of conventional biomass (9.0–12.0 MJ/kg).

A continuous work involving semi-critical assisted solvent extraction using non-chlorinated hydrocarbons and/or azeotropes are currently in progress regarding to include into our research repertoire an understandable correlation between solvent/ biomass thermochemical interaction that permit the proposal and development of more advanced extractors' design.

Chapter 7

Conclusion and New Directions

The semi-critical extraction method, SmCE, carried out in the Ch-G extractor, has proved to be a reliable process to extract high-quality proteins, amino acids, vitamins, antioxidants, hydrocarbon, and lipid substrates separated from a wide number of sources, including but not limited to bovine whole milk, infant formula milk powder, cereals, peanuts, fruits, seeds, flowers, beef, white egg (ovalbumin, yolk), different food matrix, spent coffee ground, macro and micro algae, within others, in an effort to curve some of the several difficulties encountered by the research, industry, and technological community on either extraction or separation of these valuable molecules for its further production, analysis, and commercialization.

Now a days the consensus in the extraction of medicines, food supplements, animal feed, and fertilizers, aims on the use of non-toxic halogenated solvents (including the key solvent used in phytonics extraction) because the risks and collateral activity these solvents have on the extracted material among the additional work necessary to eliminate them from the solute and the cost of its discard for recycling. Although new proposed solvent-free methods appear to be promising at the laboratory scale, they need more research at the commercial scale with an aim to minimize the use of hazardous halogenated solvents.

Our approaches in this direction relay with the improvements in a solvent-assisted extraction method that is reliable, less time

Semi-Critical Assisted Extraction: Applications and Commercialization in Biotechnology, Food, and Pharmacy
Tulio Chavez-Gil

ISBN 978-981-4968-36-2 (Hardcover), 978-1-003-29124-4 (eBook)
www.jennystanford.com

consuming, and cost effective in the production of high-quality bioproducts.

The designed Ch-G extractor is a glassware devise that works at standard conditions (25°C and 1 atm) but can also be used to operate above a solvent's boiling point (like water). It offers more advantages than other glassware extractors commercialized in the market to handle similar samples and solvents. Thus, the chemical products obtained using different solvents (preferably non-chlorinated) improved the yield of, for example, algal biofuel when compared with the data published by using the Soxhlet method and in close agreement with that obtained through the semi- to automatic supercritical methods.

Our preliminary investigations found that plain hydrocarbon solvents are not enough as algae cell wall disrupters whether compared with the effectiveness achieved with synthesized azeotropes possessing mixed polarity. SmCE shows an excellent physical chemistry correlation between hydrocarbon content and calorific values in the algal oil extracted without purification by recommended esterification steps. Thus, the high quality of hydrocarbon detected gave a C/H ratio composition of $C_{15}H_{26}$ as the main empirical formula in the macro algae *Spirogyra* biocrude oil.

The specific components identified by GC–MS analytical techniques include unsaturated hydrocarbons and phenols, in addition to heterocycles containing nitrogen as well as long-chain fatty acids and amides. In addition to the high percentages of hydrocarbon and oxygen determined by elemental analysis, the content of nitrogen, sulfur, and heavy metals detected in the macro algae harvested from urban freshwater streams was low. These findings are telling about the extraction efficiency achieved with the SmCE method by taken into consideration that in the experiments carried out on the Ch-G extractor, the sample amount was in the range of 10 to 30 g instead of commonly used 1–3 g which is recommended to be the optimum amount for critical and supercritical CO_2 extraction.

A detailed spectroscopic analysis of algae oils through FTIR shows that the lipidic components contain oxygen in combination with nitrogen by forming aromatic nitrogen compounds and free fatty acids as predominant chemical structures.

The findings on the heat capacity of algae oil are not our concluded work in this area but still addressing the significant improvement on algae cell wall disruption achieved with the SmCE method to extract lipids and fatty acid methyl esters (FAME) in both large amount and high calorific value, with these preliminary results equaling and until some extent better than the published data obtained through established technologies for the extraction of oils of conventional biomass.

We are working on algae (macro, micro) extraction through semi-critical assisted solvent methods by using pure non-halogenated hydrocarbons and mixtures of other nonhazardous solvents (ionics) in an aim to conduct our research and technological approaches to understand the biphasic equilibria solvent/biomass during its thermochemical interaction.

Like the extraction methods that are carried out under critical to supercritical conditions, the SmCE method fulfills the requirements to be as profitable and reliable as the supercritical ones. Nonetheless, SmCE is carried out in the Ch-G instrument which allows for an easy operation and works at mild conditions for an either extraction or separation of valuable biological molecules, by permitting them to keep its nature activity because they do not suffer changes in its net physical chemistry characteristics, an advantage that the critical methods cannot afford due to the extreme physical conditions (temperature, pressure) necessary to their practical operation.

Nonetheless, the method will achieve interest and recognition only when chemists, biochemists, biotechnologists, food specialists, fragrance researchers, and phytochemists experience the advantages offered by a method that can be automated according to the ideas introduced by those touched by the innovation spirit.

As advised for any other extraction method, working with SmCE technology requires a clear knowledge of both the chemical properties of the substance submitted for extraction and the solvent to be used. Moreover, to work with the Ch-G extractor, the operator/analyst needs to know the critical properties of a solvent, together with its sublimation and the volume capacity of the extractor, but not about trapping.

In addition, a basic college education in chemistry or an equivalent subject is needed to perform extraction or separation of products from a wide scope of raw samples to be processed in the Ch-G apparatus. Therefore, an operator with a basic college education, who can follow instructions, has reading comprehension, and can understand standard protocols as written information, can successfully perform an extraction, separation, or can synthesize a solvent in the Ch-G compact unit.

However, for quantitative determinations that must include trace-level analysis of substances, it will require a minimum of an analytical instrumentation preparation to satisfactorily accomplish goals.

Nowadays, the network sector prepares analysts and devise operators through workshops, symposiums, and case study demonstrations, which engage them in troubleshooting problems by using the most affordable way, a diagnostic software, due to most extraction processes (e.g., ultrasonication, pulsed liquid, microwave, pressing, centrifugation, subcritical, supercritical, and phytonics) are systems designed to form in most cases as an interface by connecting the extraction devise (the extractor) with an electronic regulator (the software) by doing of these technologies dependent of constant supervision to prevent costly reparations.

Initially, the Ch-G apparatus was constructed in glass with the aim for an easy cleaning, rinse, and dry that allows for a quick refill with same or different sample with the obtained product collected in either the reaction flask for further column chromatography separation or as separate stages individually recovered on port 8, depending on the extraction setup and solvent involved in the process or protocol. Nonetheless, the Ch-G extractor can be constructed in other materials among glass (e.g., Stanley steel or aluminum), if moderated to high pressure or temperature is the parameter to be considered for the use of a special solvent such as occur with liquids that has low boiling point (25°C–45°C) and high vapor pressure. But the Ch-G constructed in glassware and working at semi-critical conditions, shows to be a versatile technology for the extraction of volatile molecules such as flower fragrances that are used in the preparation of infant medicines, formula foods, and cosmetics very difficult to obtain by supercritical or empiric extraction technologies.

Bibliography

1. Cohen, E. R., Cvitas, T., Frey, J. G., Holmstrom, B., Kuchitsu, K., Marquardt, R., Mills, I., Pavese, F., Quack, M., Stohner, J., Strauss, H., Takami, M., and Thor, A. J. (2007). *Quantities, Units, and Symbols in Physical Chemistry, IUPAC Green Book*, 3rd edition, RSC Publishing, ISBN: 978 0 85404 433 7.
2. Atkins, P. and De Paula, J. (2006). *Physical Chemistry*, 8th edition, pp. 246, W.H. Freeman & Co, NY., ISBN: 0-7167-8759-8.
3. Brady, J. E., Russell, J. W., and Holum, J. R. (2000). *Chemistry: Matter and Its Changes*, 3rd edition, pp. 97–99, John Wiley & Sons, NY, ISBN: 0-471-18476-4.
4. Homann, K. H. (1993). *IUPAC Commission on Physicochemical Symbols, Quantities and Units*, Abbreviated List of Quantities, Units and Symbols in Physical Chemistry, Blackwell Scientific Publications, Ltd., Oxford, UK.
5. Mortimer, R. G. (2005). *Mathematics for Physical Chemistry*, 3rd edition, pp. 57, Elsevier Inc.
6. Wikipedia. 2012, https://www.cpp.edu/~lllee/gasconstant.pdf.
7. https://www.csus.edu/indiv/m/mackj/chem142/kinetics.pdf.
8. Gupta, S., Chauhan, N., and Jain, U. (2017). Laboratory Diagnosis of HbA1c: A Review. *J. Nanomed. Res.*, **5**(4): 00120. doi: 10.15406/jnmr.2017.05.00120.
9. Nadia, M., Nazrun, A., Norazlina., M., Isa, N., Norliza., M., and Ima Nirwana, S. (2012). The anti-inflammatory, phytoestrogenic, and antioxidative role of *Labisia pumila* in prevention of postmenopausal osteoporosis. *Adv. Pharmacol. Sci.*, 2012: 1–7.
10. World Health Organization (WHO) (2003). *Traditional Medicine*, WHO, Geneva, Switzerland.

11. Gias, U. (1998). Standardization of herbal preparation. *The Independent*, **30**: 13.
12. Khan, I. A. and Smillie, T. (2012). Implementing a "quality by design" approach to assure the safety and integrity of botanical dietary supplements. *J. Natural Products*, **75**(9): 1665–1673.
13. John Watson Head of Business Development Reports and Data | web: www.reportsanddata.com.
14. Diallo, D., Hveem, B., Mahmoud, M. A., Berge, G., Paulsen, B. S., and Maiga, A. (1999). An ethnobotanical survey of herbal drugs of Gourma district, *Mali. Pharm. Biol.*, **37**(1): 80–91.
15. Kris-Etherton, P., Lefevre, M., Beecher, G., Gross, M., Keen, C., and Etherton, T. (2004). Bioactive compounds in nutrition and health research methodologies for establishing biological function: The antioxidant and anti-inflammatory effects of flavonoids on atherosclerosis. *Ann. Rev. Nutr.*, **24**: 511–538.
16. Karimi, E., Jaafar, H. Z., and Ahmad, S. (2013). Antifungal, anti-inflammatory and cytotoxicity activities of three varieties of labisia pumila benth: From microwave obtained extracts. *BMC Complement Altern. Med.*, **24**: 13–20.
17. Chua, L. S., Latiff, N. A., Lee, S. Y., Lee, C. T., Sarmidi, M. R., and Aziz, R. A. (2011). Flavonoids and phenolic acids from *Labisia pumila* (Kacip Fatimah). *Food Chem.*, **127**(3): 1186–1192.
18. Yamamoto, Y. and Gaynor, R. B. (2001). Therapeutic potential of inhibition of the NF-κB pathway in the treatment of inflammation and cancer. *J. Clin Invest.*, **107**(2): 135–142.
19. Cushnie, T. and Lamb, A. J. (2005). Antimicrobial activity of flavonoids. *Int. J. Antimicrob. Agents*, **26**(5): 343–356.
20. Shahidi, F., Janitha, P., and Wanasundara, P. (1992). Phenolic antioxidants. *Crit. Rev. Food Sci. Nutr.*, **32**(1): 67–103.
21. Wei, H., Tye, L., Bresnick, E., and Birt, D. F. (1990). Inhibitory effect of apigenin, a plant flavonoid, on epidermal ornithine decarboxylase and skin tumor promotion in mice. *Cancer Res.*, **50**(3): 499–502.
22. Migliaccio, S. and Anderson, J. (2003). Isoflavones and skeletal health: Are these molecules ready for clinical application? *Osteopor. Int.*, **14**(5): 361–368.
23. Karimi, E. and Jaafar, H. Z. (2011). HPLC and GC-MS determination of bioactive compounds in microwave obtained extracts of three varieties of *Labisia pumila* Benth. *Molecules*, **16**(8): 6791–6805.

24. Karimi, E., Jaafar, H. Z., and Ahmad, S. (2011). Phytochemical analysis and antimicrobial activities of methanolic extracts of leaf, stem and root from different varieties of *Labisa pumila* Benth. *Molecules*, **16**(6): 4438–4450.

25. Ramlan, A. A., Sivakumar, K., and Chwan, D. Y. F. (2005). Engineering aspects of herbal and phytochemical processing: A Malaysian perspective. http://eprints.utm.my/5583/ (accessed January 15, 2016).

26. Sim, C., Kumaresan, S., and Sarmidi, M. R. (2004). Mass transfer coefficients of eurycoma longifolia batch extraction process. http://eprints. utm.my/6026/.

27. Volk, T. A. (2012). *The Potential of Woody Biomass for Bioenergy, Biofuels and Bioproducts*. SUNY- ESF, Syracuse, NY Biotechnology Sympos, May 21–22, 2012.

28. Ćujić, N., Šavikin, K., Janković, T., Pljevljakušić, D., Zdunić, G., and Ibrić, S. (2016). Optimization of polyphenols extraction from dried chokeberry using maceration as traditional technique. *Food Chem.*, **194**: 135–142.

29. Albuquerque, B. R., Prieto, M. A. Barreiro. M. F, Rodrigues A., Curran, T. P., Barros, L., and Ferreira, I. C. F. R. (2017). Catechin-based extract optimization obtained from *Arbutus unedo* L. fruits using maceration/ microwave/ultrasound extraction techniques. *Ind Crops Prod.*, **95**: 404–415.

30. Jovanović., A. A., Đorđević., V. B, Zdunić., G. M, Pljevljakušić., D. S., Šavikin., K. P, Gođevac., D. M, and Bugarski., B. M. (2017). Also, an optimization of the extraction process of polyphenols from *Thymus serpyllum* L. herb using maceration, heat- and ultrasound-assisted techniques. *Sep Purif Technol.*, **179**: 369–380.

31. Jin S, Yang M., Kong Y, Yao X., Wei Z., Zu Y., and Fu, Y. (2011). Microwave-assisted extraction of flavonoids from *Cajanus cajan* leaves. *Zhongcaoyao*, **42**(11): 2235–2239.

32. Wu, X., Ge, X., Liang, S., Lv, Y., and Sun, H. (2014). A novel selective accelerated solvent extraction for effective separation and rapid simultaneous determination of six anthraquinones in tartary buckwheat and its products by UPLC–DAD. *Food Anal. Meth.*, **7**(2): 1–9.

33. Laghari, A. Q., Memon, S., Nelofar, A., and Laghari, A. H. (2011). Extraction, identification and antioxidative properties of the flavonoid rich fractions from leaves and flowers of *Cassia angustifolia*. *Amer. J. Anal. Chem.*, **2**(8): 871–878.

34. Zhu, H., Wang, Y., Liu, Y., Xia Y., and Tang, T. (2010). Analysis of flavonoids in *Portulaca oleracea* L. by UV–Vis spectrophotometry with comparative study on different extraction technologies. *Food Anal. Meth.*, **3**(2): 90–97.

35. Petrucci, R H., Herring F. G., Madura J. D., and Bissonnette. C. (2011). *General Chemistry: Principles and Modern Applications,* 10th edition, Pearson Education, Upper Saddle River, New Jersey.

36. Nathan H. D. and Henrickson, C. (2001). *Chemistry*, pp. 84–86, Wiley, New York, Library of U.S.A. Congress Control Number: 2001024132., ISBN: 0-7645-6377-7.

37. Wells, M. J. M. (2003). *Sample Preparation Techniques in Analytical Chemistry*, edited by Somenath Mitra, John Wiley & Sons, Inc., pp. 38, ISBN 0-471-32845-6.

38. Mackay, D., Shiu, W. Y., and Ma, K. C. (1993). *Illustrated Handbook of Physical-Chemical Properties and Environmental Fate for Organic Chemicals, Vol. III, Volatile Organic Chemicals*, Lewis Publishers, CRC Press, Boca Raton, FL.

39. Lyman, W. J. (1982). Solubility in water, in *Handbook of Chemical Property Estimation Methods: Environmental Behavior of Organic Compounds*, edited by W. J. Lyman, W. F. Reehl, and D. H. Rosenblatt, McGraw-Hill, New York, pp. 2-1.

40. Verschueren, K. (1996). *Handbook of Environmental Data on Organic Chemicals*, 3rd edition, pp. 4–6, 20, 22, Van Nostrand Reinhold, New York.

41. Ney, R. E. (1990). *Where Did That Chemical Go? A Practical Guide to Chemical Fate and Transport in the Environment*, pp. 10, 13, 18, 32, Van Nostrand Reinhold, New York.

42. Mackay, D. and Yuen, T. K. (1980). Volatilization rates of organic contaminants from rivers. *Water Pollut. Res. J. Can.*, **15**(2): 83.

43. Mackay, D., Shiu, W. Y., and Ma, K. C. (2000). Henry's law constant, in *Handbook of Property Estimation Methods for Chemicals: Environmental and Health Sciences*, edited by R. S. Boethling and D. Mackay, pp. 69, CRC Press, Boca Raton, FL.

44. Bi, S. and Yin, D. (2000). *Plant Polyphenol*, pp. 1–46, Science Press, Beijing.

45. Dranca, F. and Oroian, M. (2018). Extraction, purification and characterization of pectin from alternative sources with potential technological applications. *Food Research International*, **113**: 327–350.

46. Khajenoori, M. and Asl, A. H. (2013). Subcritical water extraction, in *Mass Transfer: Advances in Sustainable Energy and Environment Oriented Numerical Modeling*, pp. 459–487, InTechOpen.

47. Miller, D. J. and Hawthorne, S. B. (2000). Solubility of liquid organics of environmental interest in subcritical (hot/liquid) water from 298 K to 473 K, *J. Chem. Eng. Data*, **45**: 78.

48. Ayala, R. S. and De Castro, M. L. (2001). Continuous subcritical water extraction as a useful tool for isolation of edible essential oils. *Food Chem.*, **75**(1): 109–113.

49. Ozel, M. Z., Gogus, F., and Lewis, A. C. (2003). Subcritical water extraction of essential oils from *Thymbra spicata*. *Food Chem.*, **82**(3): 381–386.

50. Shantanu, C. (2016). Stevia Technology: New Extraction Methods. https://www.steviashantanu.com/new-extraction-methods.

51. Khajenoori, M. A., Asl, H. F., Hormozi, M. H., and Eikani, H. N. (2009). Subcritical water extraction of Zataria Multiflora Boiss. *J. Food Process Eng.*, **32**: 804–816.

52. Eikani, M. H., Golmohammad, F., and Rowshanzamir, S. (2007). Subcritical water extraction of essential oils from coriander seeds (*Coriandrum sativum* L.). *J. Food Eng.*, **80**(2): 735–740.

53. Jimenez-Carmona, M., Ubera, J., and de Castro, M. L. (1999). Comparison of continuous subcritical water extraction and hydrodistillation of marjoram essential oil. *J. Chromatogr. A*, **855**(2): 625–632.

54. Saim, N., Osman, R., Yasin, W. A. H. M., and Hamid, R. D. (2008). Subcritical water extraction of essential oil from coriander (*Coriandrum sativum* L.) seeds. *Malays. J. Anal. Sci.*, **12**(1): 35–41.

55. Badrkhani, N., Khajenoori., M., and Asl, A. H. (2013). Subcritical water extraction of Thymol from Ajowan seeds by Taguchi experimental design., *J. Appl. Chem.*, **7**(25): 51–58.

56. Grosso, C., Cardoso, M. T., Figueiredo, A., Moldão-Martins, M., Burillo, J., Urieta, J., Barroso, J., Coelho, J., and Palavra, A. (2007). Supercritical fluid extraction, Hydrodistillation and Soxhlet extraction of the aerial part of winter savory. In *Proceedings 528 of European Congress of Chemical Engineering (ECCE-6)*, Copenhagen, 16–20 September 2007.

57. Gao, X., Yani, S., and Wu, H. (2014). Pyrolysis of spent biomass from mallee leaf steam distillation: Biochar properties and recycling of inherent inorganic nutrients. *Energy Fuels*, **28**(7): 4642–4649.

58. Alders, L. (1959). *Liquid–Liquid Extraction*, 2nd edition, Elsevier, Amsterdam.

59. Johnson, L. A. and Lusas, E. W. (1983). Comparison of alternative solvents for oils extraction. *J. Am. Oil Chem. Soc.*, **60**: 229–242. doi: 10.1007/BF02543490.

60. Akgün, N. A., Bulut, H., Kikic, I., and Solinas, D. (2014). Extraction behavior of lipids obtained from spent coffee grounds using supercritical carbon dioxide. *Chem. Eng. Technol.*, **37**: 1975–1981. doi: 10.1002/ceat.201400237.

61. Wilkinson, A. M. and McNaught, A. D. (1997). Partition coefficient, in *Compendium of Chemical Terminology: IUPAC Recommendations*, edited by M. Nič, J. Jirát, B Košata, A. Jenkins, and A. McNaught, Blackwell Science, Oxford. doi: 10.1351/gold book. ISBN 978-0-86542-684-9.

62. Foster, F., Cabrices, O, Whitecavage, J., Stuff, J., and Pfannkoch, E. (2014). Automating liquid-liquid extractions using a bench-top workstation. *Gerstel, Inc., 701 Digital*. Linthicum, MD, USA. www.gerstel.com/pdf/p-lc-an-2014-07.pdf.

63. Armarego, W. L. F. and Lin-Chai, C. L. (2009). *Purification of Laboratory Chemicals*, 6th edition, pp. 7–9, Elsevier, Oxford, UK. ISBN: 978-1-85617-567-8.

64. Kister, H. Z. (1992). *Distillation Design*, 1st edition, *McGraw-Hill*, ISBN 0-07-034909-6.

65. Perry, A. B. and Green, R. H. (Editors) (1997). *Perry's Chemical Engineers' Handbook*, 7th edition, McGraw-Hill, ISBN 0-07-049841-5.

66. Griffith, A. P. and Collison, M W. (2001). Improved methods for the extraction and analysis of isoflavones from soy-containing foods and nutritional supplements by reversed-phase high-performance liquid chromatography and liquid chromatography-mass spectrometry. *J Chromatogr A*. **913**(1-2): 397–413.

67. Seader, J. D. and Henley, E. J. (1998). *Separation Process Principles*, Wiley, NY, ISBN 0-471-58626-9.

68. DePriester, C. L. (1953). Light-hydrocarbon vapor-liquid distribution coefficients. *Chem. Eng. Prog, Symposium Series*, **7**(49): 1–43.

69. Handley, A. J. (1999). *Extraction Methods in Organic Analysis*, Shef Reld Academic Press, Shef Reld.

70. Andrasi, N., Helenkar, A., Vasanits-Zsigrai, A., Zaray, G., and Molnar-Perl, I. (2011). The role of the acquisition methods in the analysis of natural and synthetic steroids and cholic acids by gas chromatography-mass spectrometry, *J. Chromatogr. A*, 1218: 8264–8272.

71. Arditsoglou, A. and Voutsa. (2008). Determination of phenolic and steroid endocrine disrupting compounds in environmental matrices, *Environ. Sci. Pollut. Res.*, **15**: 228–236.

72. Patel, K., Panchal, N., and Ingle, P. (2019). Extraction methods: Microwave, ultrasonic, pressurized fluid, soxhlet extraction, etc. *Inter. J. Adv. Res. Chem. Sci.*, **6**(3): 6–21.

73. Darwesh, O. M. (2014). Extraction of bio-active compounds extracted from inula helenium roots by leaching process. *Res. J. Pharm. Biol. Chem. Sci.*, **5**(6): 254–261.

74. Mostoufi, N., Faridkhou, A., Gharebagh, R. S., and Norouzi, H. R. (2010). Dynamic modelling of the sugar extraction process from sugar beet. *International Food Information Service*, **3**(1): 49–56. doi: 10.1616/1750-2683.0052.

75. Naima, R., Hannache, H., Oumam, M., Sesbou, A., Charrier, B., Pizzi, A., and Bouhtoury, F. C.-E. (2015). Green extraction process of tannins obtained from Moroccan *Acacia mollissima* barks by microwave: Modeling and optimization of the process using the response surface methodology. *Arabian J. Chem.* 1-37.

76. Fajriyati, M., Meta, M., Amran, L., and Zainal, Z. (2017). Optimization of mango seed kernel oil extraction using response surface methodology. *OCL*, 24: 5.

77. Wu, C., Spongberg, A. L., Witter, J. D., Fang, M., Czajkowski, K. P., and Ames, A. (2010). Dissipation and leaching potential of selected pharmaceutically active compounds in soils amended with biosolids. *Arch. Environ. Contam. Toxicol.*, **59**: 343–351.

78. Chavez-Gil, T., Sobhi, H. F., and Pacheco-Laracuente, A. (2018). Improvements on semi critical assisted-solvent extraction of biocrude oil from *Spirogyra* sp. macro algae. *Am. J. Biochem. Biotechnol.*, **14**(2): 124–136. doi: 10.3844/ajbbsp.2018.124.136.

79. Sekine, T. and Hasegawa, Y. (1977). *Solvent Extraction Chemistry*, pp. 919, Marcel Dekker, New York.

80. Toledo, T. R. (1993). Extraction, in *Fundamentals of Food Process Engineering*, Springer, Boston, MA. doi: https://doi.org/10.1007/978-1-4615-7052-314.

81. Rao, D. G. (2009). Extraction, in *Fundamentals of Food Engineering*, pp. 364–368, PHI, New Delhi.

82. Berk, Z. (2009). Extraction, in *Food Process Engineering and Technology, Direct Sciences*, pp. 375–390, eBook ISBN 9780080920238.

83. Johnson, L. A. and Lusas, E. W. (1983). Comparison of alternative solvents for oils extraction. *J. Am. Oil Chem. Soc.*, **60**: 229–242. doi: 10.1007/BF02543490.

84. Reichardt, C. and Welton, T. (2010). Solute–solvent interactions, in *Solvents and Solvent Effects in Organic Chemistry*, pp. 7–64, Wiley-VCH Verlag GmbH & Co. KGaA.

85. Mustafa, A. and Turner, C. (2011). Pressurized liquid extraction as a green approach in food and herbal plants extraction: A review. *Anal. Chim. Acta*, **703**(1): 8–18. PMID: 21843670, doi: 10.1016/j.aca.2011.07.018

86. Lo, T. C., Baird, M. H. I., and Hanson, C. (1983). *Handbook of Solvent Extraction*, pp. 1980, Wiley-Interscience, New York.

87. Abdullah, M. and Koc, A. B. (2013). Oil removal from waste coffee grounds using two-phase solvent extraction enhanced with ultrasonication. *Renew. Energy*, **50**: 965–970. doi: 10.1016/j.renene.2012.08.073.

88. Kou, D. and Mitra, S. (2003). Extraction of semi-volatile organic compounds from solid matrices, in *Sample Preparation Techniques in Analytical Chemistry* **162**, edited by S. Mitra, pp. 139–182, John Wiley & Sons, New Jersey. ISBN 0-471-32845-6.

89. Luque de Castro, M. D. and Priego-Capote, F. (2010). Soxhlet extraction: Past and present panacea. *J. Chromatogr. A.*, **1217**(16): 2383–2389. doi: 10.1016/j.chroma.2009.11.027.

90. Randall, E. L. (1974). Improved method for fat and oil analysis by a new process of extraction. *JAOAC* **57**: 1165–1168.

91. Al-Hamamre, Z., Foerster, S., Hartmann, F., Kröger, M., and Kaltschmitt, M. (2012). Oil extracted from spent coffee grounds as a renewable source for fatty acid methyl ester manufacturing. *Fuel* **96**: 70–76. doi: 10.1016/j. fuel.2012.01.023.

92. Shirley, A. (2004). Soxtec: Its principles and applications, in *Oil Extraction and Analysis: Critical Issues and Comparative Studies*, AOCS Publishing.

93. Thiex, N., Anderson, S., and Gildemeister, B. (2003). Crude fat, hexanes extraction, in feed, cereal grain, and forage (Randall/Soxtec/Submersion method): A collaborative study, *JAOAC Int.*, **86**: 888–898.

94. Heinrich, C. (1970). Use of the Rose–Gottlieb method for rapid gravimetric fat determination with the Heraeus apparatus. *Deutsche Milchwirtschaft.*, **21**(20): 797–799.

95. Thiex, N., Anderson, S., and Gildemeister, B. (2003). Crude fat, diethyl ether extraction, in feed, cereal grain, and forage (Randall/ Soxtec/Submersion method): A collaborative study, *JAOAC Int.*, **86**: 899–908.

96. Abeysena, I. and Darrington, R. (2013). Understanding Evaporation and Concentration Technologies Part 1 – Basic Principles of Commonly Used Evaporation Technologies. American Laboratory website. https://www.americanlaboratory.com/914-Application-Notes/130187-.

97. Armarego, W. L. and Lin-Chai, C. L. (2009). *Purification of Laboratory Chemicals*, 6th edition, pp. 17, Butterworth-Heinemann, Burlington, MA, USA.

98. Rotavapor (BÜCHI - Rotavapor® R-100). Adapted with permission of BÜCHI. Co.

99. Matsumoto, M., Otono, T., and Kondo, K. (2001). Synergistic extraction of organic acids with tri-n-octylamine and tri-n-butylphosphate. *Sep. Purif. Technol.*, **24**: 337.

100. Caşcaval, D., Galaction, A. I., and Kloetzer, L. (2012). Synergic extraction of folic acid with di(2-ethylhexyl) phosphoric acid and amberlite LA-2. *Sep. Sci. Technol.*, **47**: 834.

101. Keshav, A., Wasewar, K. L., Chand, S., and Uslu, H. (2009). Effect of binary extractants and modifier–diluents systems on equilibria of propionic acid extraction. *Fluid Phase Equilibr.*, **275**: 21.

102. Miazek, K., Kratky, L., and Sulc, R. (2017). Effect of organic solvents on microalgae growth, metabolism and industrial bioproduct extraction: A review. *Int. J. Mol. Sci.*, **18**(7), Open Access Article ID 9209210 | https://doi.org/10.1155/2019/9209210.

103. Hong, Y. K. and Hong, W. H. (2000). Extraction of succinic acid with 1-octanol/n-heptan solutions of mixed tertiary amines. *Bioproc. Biosyst. Eng.*, **23**: 535.

104. Hong, Y. K. and Hong, W. H. (1999). Reactive extraction of lactic acid with mixed tertiary amine extractants. *Biotechnol. Techniq.*, **13**: 915.

105. Kroschwitz, J. I. and Kirk-Othmer, M. (2001). *Encyclopedia of Chemical Technology*, 4th edition, John Wiley & Sons, Inc., London.

106. Kivrak, I. (2015). Chemical constituents: Water-soluble vitamins, free amino acids, and sugar profile from *Ganoderma adspersum*. *Nat Prod Res.*, **28, 29**(6): 518–523.

107. Poştaru, M., Bompa, A. S., Galaction, A. I., Blaga, A. C., and Caşcaval, D. (2016). Comparative study on pantothenic acid separation by reactive extraction with tri-n-octylamine and di-(2-ethylhexyl) phosphoric acid. *Chem. Biochem. Eng. Quart.*, **30**(1): 81–92.

108. Bai, X., Ghasemi, N. F., Ye, L., Lant, P., and Pratt, S. (2014). Enhanced lipid extraction from algae using free nitrous acid pretreatment. *Bioresour. Technol.*, **159**: pp. 36–40.

109. Kim, Y. H., Choi, Y. K., Park, J., Lee, S., Yang, Y. H., Kim, H. J., Tae-Joon Park, T. J., Kim, Y. H., and Lee, S. H. (2012). Ionic liquid-mediated extraction of lipids from algal biomass. *Bioresour. Technol.*, **109**: 312–315.

110. Boyd, A. R., Champagne, P., McGinn, P. J., MacDougall, K. M., Melanson, J. E., and Jessop, P. G. (2012). Switchable hydrophilicity solvents for lipid

extraction from microalgae for biofuel production. *Bioresour. Technol.*, **118**: 628–632.

111. Zhou, W., Wang, Z., Asraful, A M., Xu J., Zhu, S., Yuan, Z., Huo, S., Guo, Y., Qin, L., and Ma, L. (2019). Repeated utilization of ionic liquid to extract lipid from algal biomass. *Int. J. Polym. Sci.*, Open Access Article. ID 9209210, 7 pages https://doi.org/10.1155/2019/9209210.

112. Kim, Y.-H., Choi, Y.-K., and Park, J. (2012). Ionic liquid-mediated extraction of lipids from algal biomass. *Bioresour. Technol.*, **109**: 312–315.

113. Abimbola, T., Christodoulatous, C., and Adeniyi, L. A. (2021). Performance and optimization studies of oil extraction from *Annochloropsis* spp. and *Scenedesmus obliquus*. *J. Cleaner Product.*, **11**(8): 127295. doi: 10.1016/j.jclepro.2021.127295.

114. King, J. W., Snyder, J. M., Taylor, S. L., and Johnson, J. H. (1993). Translation and optimization of supercritical fluid extraction methods to commercial instrumentation. *J. Chromatogr. Sci.*, **31**: 1–5.

115. Galaction, A. I., Bompa, A. S., Kloetzer, L., Turnea, M. A., and Caçaval, D. (2015). Synergistic extraction and transport of folic acid through liquid membranes. *Solvent Extr. Ion Exch.*, **33**: 313–328.

116. Gabaldón, J. A., Maquieira, A., and Puchades, R. (1999). Current trends in immunoassay-based kits for pesticide analysis. *Crit. Rev. Food Sci. Nutr.*, **39**(6): 519–538.

117. Suslick, K. S. and Flannigan, D. J. (2008). Inside a collapsing bubble: Sonoluminescence and the conditions during cavitation. *Ann. Rev. Phys. Chem.*, **59**: 659–683.

118. Khanal, S. K., Grewell, D., Sung, S., and Van Leeuwen, J. (2007). Ultrasound applications in wastewater sludge pretreatment: A review. *Crit. Rev. Environ. Sci. Technol.*, **37**(4): 273–313.

119. Chuyen, H. V., Nguyen, M. H., Roach, P. D., Golding, J. B., and Parks, S. E. (2018). Microwave-assisted extraction and ultrasound-assisted extraction for recovering carotenoids from Gac peel and their effects on antioxidant capacity of the extracts. *Food Sci. Nut.*, **6**: 189–196.

120. Hosikian, A., Lim, S., Halim, R., and Danquah, M. K. (2010). Chlorophyll extraction from microalgae: A review on the process engineering aspects. *Int. J. Chem. Eng.*, 1–11. Article ID 391632, doi: 10.1155/2010/391632.

121. Ordóñez-Santos, L. E., Pinzón-Zarate, L. X., and González-Salcedo, L. O. (2015). Optimization of ultrasonic-assisted extraction of total carotenoids from peach palm fruit (*Bactris gasipaes*) by-products

with sunflower oil using response surface methodology. *Ultrason. Sonochem.*, **27**: 560–566.

122. Raso, J. and Heinz, V. (Editors). (2006). Pulsed electric field technology for the food industry, in *Fundamentals and Applications*, Springer, New York.

123. Mohamed, M. E. A., Ayman, H., and Eissa, A. (2012). *Pulsed Electric Fields for Food Processing Technology*, InTechOpen.

124. Mohamed, M. E. A. and Amer Eissa, A. H. (2012). Structure and function of food engineering, in *Pulsed Electric Fields for Food Processing Technology*, edited by H. Ayman and A. Eissa, pp. 275–306, InTechOpen, doi: 10.5772/48678.

125. Wang, L. and Weller, C. L. (2006). Recent advances in extraction of nutraceuticals from plants. *Trends Food Sci. Technol.*, **17**: 300–312.

126. Kawamura, F., Kikuchi, Y., Ohira, T., and Ytagai, M. (1999). Accelerated solvent extraction of paclitaxel and relate compounds from the bark of Taxus cuspidate. *J. Nat. Prod.*, **62**: 244–247.

127. Richter, B. E. (1997). Extraction of polychlorinated dibenzo-p-dioxins and polychlorinated dibenzofurans from environmental samples using accelerated solvent extraction (ASE). *Chemosphere*, **34**(5–7): 975–987.

128. Alvarez-Rivera, G., Mendiola, J. A., Bueno, M., and Ibáñez, E. (2020). Pressurized Liquid Extraction, in *Liquid Phase Extraction*, pp. 375–398. ISBN- 978-0-12-816911.

129. Andressa, K., Eduardo, J., Cardozo-filho, L., and Antonio, E. (2019). Pressurized liquid and ultrasound-assisted extraction of α-bisabolol from candeia (*Eremanthus erythropappus*) wood. *Ind. Crop. Prod.*, **130**: 428–435.

130. Garcia-Galan, M. J., Diaz-Cruz, S., and Barcelo, D. (2013). Multiresidue trace analysis of sulfonamide antibiotics and their metabolites in soils and sewage sludge by pressurized liquid extraction followed by liquid chromatography-electrospray-quadrupole linear ion trap mass spectrometry. *J. Chromatogr. A*, **1275**: 32–40. doi: 10.1016/j.chroma.2012.12.004.

131. Wang, N., Su, M., Liang, S., and Sun, H. (2016). Sensitive residue analysis of quinolones and sulfonamides in aquatic product by capillary zone electrophoresis using large-volume sample stacking with polarity switching combined with accelerated solvent extraction. *Food Anal. Methods*, **9**(4): 1020–1028.

132. Li, D., Zhang, K., Chen, L., Ding, M., Zhao, M., and Chen, S. (2017). Selection of *Schizochytrium limacinum* mutants based on butanol tolerance. *Electron. J. Biotechnol.*, **30**: 58–63.

133. Kadam, S., Tiwari, B., and O'Donnell, C. (2013). Application of novel extraction technologies for bioactives from marine algae. *J. Agric. Food Chem.*, **61**: 4667–4675.

134. Pacheco-Laracuente, A., Villarin-Ayala, S, and Chavez-Gil, T. (2014). High performance of algae-oil extraction: The effect of azeotropic mixtures on cell wall penetration. Procedures 247th *ACS-Spring meeting*. Dallas, TX. DIVISION: Energy and Fuels. Paper # 283.

135. Bligh, E. and Dyer W. (1959). A rapid method of total lipid extraction and purification. *Can. J. Biochem. Physiol.*, **37**: 911–917.

136. Rodrigues, E., Pardal, M., Salgueiro-González, N., Muniategui-Lorenzo, S., and Alpendurada, M. (2016). A single-step pesticide extraction and clean-up multi-residue analytical method by selective pressurized liquid extraction followed by on-line solid phase extraction and ultra-high-performance liquid chromatography-tandem mass spectrometry for complex matrices. *J. Chromatogr. A*, **1452**: 10–17.

137. Anaëlle, T., Serrano Leon, E., Laurent, V., Ibanez, E., Mendiola, J., Stéphane, C., Nelly, K., La Barre, S., Luc, M., and Stiger-Pouvreau, V. (2013). Green improved processes to extract bioactive phenolic compounds from brown macroalgae using *Sargassum muticum* as model. *Talanta*, **104**: 44–52.

138. Otero, P., Quintana, S. E., Reglero, G., Fornari, T., and García-Risco, M. R. (2018). Pressurized liquid extraction (PLE) as an innovative green technology for the effective enrichment of galician algae extracts with high quality fatty acids and antimicrobial and antioxidant properties. *Mar. Drugs.*, **16**: 156. doi: 10.3390/md16050156.

139. Castuma, J. C., Catala, A., and Brenner, R. R. (1972). Oxidative desaturation of eicosa-8, ll -dienoic acid to eicosa-5,8,1 I -trienoic acid: comparison of different diets on oxidative desaturation at the 5,6 and 6,7 positions. *J. Lipid Res.*, **13**: 783–789.

140. Catalá, A. (2013). Five decades with polyunsaturated fatty acids: Chemical synthesis, enzymatic formation, lipid peroxidation and its biological effects. *J. Lipids.*, 1–9. An Open access article.

141. Castejón, N., Luna, P., and Señoráns, F. (2017). Ultrasonic removal of mucilage for pressurized liquid extraction of omega-3 rich oil from chia seeds (*Salvia hispanica* L.). *J. Agric. Food Chem.*, **65**, 2572–2579.

142. Oliveira, R., Oliveira, V., Aracava, K., and da Costa Rodrigues, C. (2012). Effects of the extraction conditions on the yield and composition of rice bran oil extracted with ethanol: A response surface approach. *Food Bioprod. Process.*, **90**: 22–31.

143. Mazin, N. (2004). Solvent extraction method and apparatus. WIPO Patent Application: WO/2004/026430.

144. Sánchez-Camargo, A., Pleite, N., Mendiola, J., Cifuentes, A., Herrero, M., Gilbert-López, B., and Ibanez, E. (2018). Development of green extraction processes for *Nannochloropsis gaditana* biomass valorization. *Electrophoresis*, doi: 10.1002/elps.201800122.

145. Gilbert-López, B., Barranco, A., Herrero, M., Cifuentes, A., and Ibáñez, E. (2017). Development of new green processes for the recovery of bioactives from *Phaeodactylum tricornutum*. *Food Res. Int.*, **99**: 1056–1065.

146. Paul, A., Frederich, M., Megido, R., Alabi, T., Malik, P., Uyttenbroeck, R., Francis, F., Blecker, C., and Haubruge, E. (2017). Insect fatty acids: A comparison of lipids from three Orthopterans and *Tenebrio molitor* L. larvae. *J. Asia-Pac. Entomol.*, **20**: 337–340.

147. Shah, M. V. and Rohit, M. C. (2013). Novel techniques for isolation and extraction of phyto-constituents from herbal plants. *Am. J. Phytomedi. Clin Therapeut.*, **3**: 338–350.

148. Margaris, N., Koedam, A., and Vokou, D. (1982). *Aromatic Plants: Basic and Applied Aspects*, The Hague/Martinus Nijhoff Publishers, London, Boston.

149. Surburg, H. and Panten, J. (2006). *Common Fragrance and Flavor Materials: Preparation, Properties and Uses*, 5th edition, Wiley-VCH, Weinheim.

150. Sørensen, L., Silva, M. S., Meier, S., and Booth, A. M. (2015). Advances in miniaturization and increasing sensitivity in analysis of organic contaminants in marine biota samples. *Trends Environ. Anal. Chem.*, **6–7**: 39–47.

151. Ruppert, L., Lin. Z.-Q., Dixon, R. P., and Johnson, K. A. (2013). Assessment of solid phase microfiber extraction fibers for the monitoring of volatile organo arsenicals emitted from a plant–soil system. *J. Hazard Mater.*, **262**: 1230–1236.

152. Analytical Chemistry 2.0. Chapter 7: "Obtaining and Preparing Samples forAnalysis". http://dpuadweb.depauw.edu/harvey_web/eTextProject/version_2.0.html

153. Waters Corporation Solid Phase Extraction Online Tutorial http://www.waters.com/waters/en_US/SPE-Method-Development/nav.htm?cid=10083845.

154. Agilent Application Note: Sasso, Pat "PAHs in Chocolate and Peanuts with Agilent J & W Select PAH and Longer GC Columns" http://www.agilent.com/cs/library/applications/5991-2299EN.pdf.

155. Sigma Aldrich Bulletin 910 "Guide to Solid Phase Extraction" http://www.sigmaaldrich.com/Graphics/Supelco/objects/4600/4538.pdf.

156. Sengers, J. V. and Levelt-Sengers, J. M. H. (1986). Thermodynamic behavior of fluids near the critical point. *Ann. Rev. Phys. Chem.*, **37**: 189. doi: 10.1146/annurev.pc.37.100186.001201.

157. Engineering ToolBox, (2018). Carbon Dioxide - Thermophysical Properties. [online] Available at: https://www.engineeringtoolbox.com/CO2-carbon-dioxide-properties-d_2017.html

158. Ridduck, J. A. and Bunger, W. B. (1970). *Organic Solvents: Physical Properties and Methods of Purification*, 3rd edition, Wiley-Interscience, New York, USA.

159. Wilhoit, R. C., Chao, J., and Hall, K. R. (1985). Thermodynamic properties of key organic compounds in the carbon range C1 to C4. Part 1. Properties of condensed phases. *J. Phys. Chem. Ref.*, **14**: 1.

160. www.vaxasoftware.com.

161. www.WikiMedia Commons; http://manualslibrary.com

162. Bouallegue, K., Allaf, T., Ben Younes, R., Téllez-Pérez, C., Besombes, C., and Allaf, K. (2020). Pressure, temperature and processing time in enhancing *Camelina sativa* oil extraction by instant controlled pressure-drop (DIC) texturing pre-treatment. *Grasas Aceites.*, **71**(3): e365. Available from:https://grasasyaceites.revistas.csic.es/index.php/grasasyaceites/article/view/1832.

163. Mkaouar, S., Bahloul, N., Gelicus, A., Allaf, K., and Kechaou, N. (2015). Instant controlled pressure drop texturing for intensifying ethanol solvent extraction of olive (*Olea europaea*) leaf polyphenols. *Sep. Purif. Technol.*, **145**: 139–146. doi: 10.1016/j.seppur.2015.03.014.

164. Chemat, F., Vian, M. A., and Cravotto, G. (2012). Green extraction of natural products: Concept and principles. *Int. J. Mol. Sci.*, **13**: 8615–8627.

165. Allaf, K., Besombes, C., Berka, B., Kristiawan, M., Sobolik, V., and Allaf, T., (2011). *Enhancing Extraction Processes in the Food Industry*, pp. 255–302, CRC Press, USA.

166. Chemat, F., Rombaut, N., Meullemiestre, A., Turk, M., Perino, S., Fabiano-Tixier, A. S., and Abert-Vian, M. (2017). Review of Green Food Processing techniques. Preservation, transformation, and extraction. *Innovative Food Sci. Emerging Technol.*, **41**: 357–377.

167. Sanchez-Valdepenas, V., Barrajon, E., Vegara, S., Funes, L., Marti, N., Valero, M., and Saura, D. (2015). Effect of instant controlled pressure drop (DIC) pre-treatment on conventional solvent extraction of

phenolic compounds from grape stalk powder. *Ind. Crops Prod.*, **76**: 545–549.

168. Allaf, T., Fine, F., Tomao, V., Nguyen, C., Ginies, C., and Chemat, F. (2014). Impact of instant controlled pressure drop pre-treatment on solvent extraction of edible oil from rapeseed seeds. *OCL*, **21**(3): A301.

169. Eikani, M. H., Khandan, N., and Feyzi, E. (2012). Instant controlled pressure drop (DIC) effect on compositional analysis of olive leaves (*Olea europaea* L.). *Food Bioprod. Process.*, **117**: 241–249.

170. Rombaut, N., Tixier, A. S., Billy A., and Chemat, F. (2014). Green extraction processes of natural products as tools for biorefinery. *Biofuels, Bioprod. Biorefin.*, **8**: 530–544.

171. Allaf, T. and Allaf, K. (2014). *Instant Controlled Pressure Drop (D.I.C.) in Food Processing*, Food Engineering Series, Springer, New York.

172. Chemat, F., Abert Vian, M., Ravi, H. K., Khadhraoui, B., Hilali, S., Perino, S., Fabiano, and Tixier, A. S. (2019). Review of alternative solvents for green extraction of food and natural products: Panorama, principles, applications and prospects. *Molecules*, **24**: 3007.

173. da Silva, C. M., Zanqui, A. B., Gohara, A. K., de Souza, A. H., Cardozo-Filho, L., Visentainer, J. V., Rovigatti, C L., Bittencourt, P. R., da Silva, E. A., and Matsushita, M. (2015). Compressed n-propane extraction of lipids and bioactive compounds from perilla (*Perilla frutescens*). *J. Supercrit. Fluids*, **102**: 1–8.

174. Corso, M. P., Fagundes-Klen, M. R., Silva, E. A., Cardozo, L., Santos, J. N., Freitas, L. S., and Dariva, C. (2010). Extraction of sesame seed (*Sesamun indicum* L.) oil using compressed propane and supercritical carbon dioxide. *J. Supercrit. Fluids*, **52**: 56–61.

175. Kanda, H., Li, P., Goto, M., and Makino, H. (2015). Energy-saving lipid extraction from wet *Euglena gracilis* by the low-boiling-point solvent dimethyl ether. *Energies*, **8**: 610–620.

176. Kanda, H., Li, P., Yoshimura, T., and Okada, S. (2013). Wet extraction of hydrocarbons from *Botryococcus braunii* by dimethyl ether as compared with dry extraction by hexane. *Fuel*, **105**: 535–539.

177. Zanqui, A. B., de Morais, D. R., da Silva, C. M., Santos, J. M., Chiavelli, U. R. L., Bittencourt, P. R. S., Eberlin, M. N., Visentainer, J. V., Cardozo-Filho, L., and Matsushita, M. (2015). Subcritical extraction of *Salvia hispanica* L. oil with n-propane: Composition, purity and oxidation stability as compared to the oils obtained by conventional solvent extraction methods. *J. Braz. Chem. Soc.*, **26**(2): 282–289.

178. Goto, M., Kanda, H., and Machmudah, S. (2015). Extraction of carotenoids and lipids from algae by supercritical CO_2 and subcritical dimethyl ether. *J. Supercrit. Fluids*, **96**: 245–251.

179. Hoshino, R., Ogawa, M., Murakami, K., Kanda, H., and Goto, M. (2017). Extraction of lipids from wet *Arthrospira platensis* by liquefied dimethyl ether. *Solvent Extr. Res. Dev.*, **24**: 47–60.

180. Bier, M. C. J., Medeiros, A. B., de Oliveira, J. S., Côcco, L. C., da Luz Costa, J., de Carvalho, J. C., and Soccol, C. R. (2016). Liquefied gas extraction: A new method for the recovery of terpenoids from agroindustrial and forest wastes. *J. Supercrit. Fluids*, **110**: 97–102.

181. Hoshino, R., Wahyudiono, Machmudah, S., Kanda, H., and Goto, M. (2014). Simultaneous extraction of water and essential oils from citrus leaves and peels using liquefied dimethyl ether. *J. Nutr. Food Sci.*, **4**: 301. doi: 10.4172/2155-9600.1000301

182. Novello, Z., Scapinello, J., Magro, J. D., Zin, G., Luccio, M. D., Tres, M. V., and Oliveira, J. V. (2015). Extraction, chemical characterization and antioxidant activity of andiroba seeds oil obtained from pressurized n-butane. *Crops Prod.*, **76**: 697–701.

183. Rapinel, V., Rombaut, N., Rakotomanomana, N., Vallageas, A., Cravotto, G., and Chemat, F. (2017). An original approach for lipophilic natural products extraction: Use of liquefied n-butane as alternative solvent to n-hexane. *LWT - Food Sci. Technol*, **85**: 524–533.

184. Sapkale, G. N., Patil, S. M., Surwase, U. S., and Bhatbhage, P. K. (2010). Supercritical fluid extraction review. *Int. J. Chem. Sci*, **8**: 729–743.

185. Khaw, K-Y., Parat, M-O., Shaw, P. N., and Falconer, J. R. (2017). Solvent supercritical fluid technologies to extract bioactive compounds from natural sources: A review. *Molecules*, **22**: 1186. doi: 10.3390/molecules22071186.

186. Brazhkin, V. V., Lyapin, A. G., Ryzhov, V. N., Trachenko, K., Fomin, D. Y., and Tsiok, E. N. (2012). Where is the supercritical fluid on the phase diagram? *Phys. Usp.*, **55**: 1061–1079.

187. Griessen, R., Boelsma, C., Schreuders, H., Broedersz, C. P., Gremaud, R., and Dam, B. (2020). Single quality factor for enthalpy-entropy compensation, isoequilibrium and isokinetic relationships. *Chem. Phys. Chem*, **21**: 1632–1643.

188. Uwineza, P A. and Waśkiewicz, A. (2020). Recent advances in supercritical fluid extraction of natural bioactive compounds from natural plant materials. *Molecules*, **25**: 3847. doi: 10.3390/molecules25173847.

189. Raventós, M., Duarte, S., and Alarcón, R. (2002). Application and possibilities of supercritical CO_2 extraction in food processing industry: An overview. *Food Sci. Technol. Int.*, **8**: 269–284.

190. Schuster, J. J., Bahr, L A., Fehr, L., Tippelt, M., Schulmeyr, J., Wuzik, A., and Braeuer, A. S. (2018). Online monitoring of the supercritical CO_2 extraction of hop. *J. Supercrit. Fluids*, **133**: 139–145.

191. Putra, N R., Yian, L N., Nasir, H M., Idham, Z B., and Yunus, M A C. (2018). Effects of process parameters on peanut skins extract and CO_2 diffusivity by supercritical fluid extraction. *IOP Conf. Ser. Mater. Sci. Eng.*, **334**: 1–10.

192. Bubalo, C. M., Vidović, S., Redovniković, I. R., and Jokić, S. (2018). New perspective in extraction of plant biologically active compounds by green solvents. *Food Bioprod. Process*, **109**: 52–73.

193. Espinosa-Pardo, A. F., Nakajima, M. V., Macedo, A. G., Macedo, A. J., and Martínez, J. (2017). Extraction of phenolic compounds from dry and fermented orange pomace using supercritical CO_2 and co-solvents. *Food Bioprod. Process*, **101**, 1–10.

194. Pourmortazavi, M. S. and Hajimirsadeghi, S. S. (2007). Supercritical fluid extraction in plant essential and volatile oil analysis. *J. Chromatogr. A*, **1163**: 2–24.

195. Natolino, A. and Da Porto, C. (2019). Supercritical carbon dioxide extraction of pomegranate (*Punica granatum* L.) seed oil: Kinetic modelling and solubility evaluation. *J. Supercrit. Fluids*, **151**: 30–39.

196. Mazzutti, S., Ferreira, S. R. S., Riehl, C. A. S., Smania, A., Smania, F. A., and Martinez, J. (2012). Supercritical fluid extraction of *Agaricus brasiliensis*: Antioxidant and antimicrobial activities. *J. Supercrit. Fluids*, **70**: 48–56.

197. Reverchon, E. and De Marco, I. (2006). Supercritical fluid extraction and fractionation of natural matter. *J. Supercrit. Fluids*, **38**: 146–166.

198. Sánchez-Camargo, A. P., Mendiola, A. J., Ibáñez, E., and Herrero, M. (2014). Supercritical fluid extraction, in *Reference Module in Chemistry, Molecular Sciences and Chemical Engineering*, edited by J. Reedijk and M. A. Waltham, Elsevier, Amsterdam, The Netherlands.

199. Bimakr, M., Rahman, R. A., Taip, F. S., Ganjloo, A., Salleh, L. M., Selamat, J., Hamid, A., and Zaidul, I. S. M. (2011). Comparison of different extraction methods for the extraction of major bioactive flavonoid compounds from spearmint (*Mentha spicata* L.) leaves. *Food Bioprod. Process*, **89**: 67–72.

200. Woźniak, Ł., Marszałek, K., Skąpska, S., and Jędrzejczak, R. (2017). The application of supercritical carbon dioxide and ethanol for the extraction of phenolic compounds from chokeberry pomace. *Appl. Sci.*, **7**: 322.

201. Akay, S., Alpak, I., and Yesil-Celiktas, O. (2011). Effects of process parameters on supercritical CO_2 extraction of total phenols from strawberry (*Arbutus unedo* L.) fruits: An optimization study. *J. Sep. Sci.*, **34**, 1925–1931.

202. Shi, J., Yi, C., Ye, X., Xue, S., Jiang, Y., and Liu, D. (2010). Effects of supercritical CO_2 fluid parameters on chemical composition and yield of carotenoids extracted from pumpkin. *LWT Food Sci. Technol.*, **43**: 39–44.

203. Rodrigues, V. M., Rosa, P. T. V., Marques, M. O. M., Petenate, A. J., and Meireles, M. A. A. (2003). Supercritical extraction of essential oil from aniseed (*Pimpinella anisum* L.) using CO_2: Solubility, kinetics, and composition data. *J. Agric. Food Chem.*, **51**, 1518–1523.

204. Azmir, J., Zaidul, I. S. M., Rahman, M. M., Sharif, K. M., Mohamed, A., Sahena, F., Jahurul, M. H. A., Ghafoor, K., Norulaini, N. A. N., and Omar, A. K. M. (2013). Techniques for extraction of bioactive compounds from plant materials: A review. *J. Food Eng.*, **117**, 426–436.

205. Sasidharan, S., Chen Y., Saravanan, D., Sundram, K. M., and Latha, L. (2011). Extraction, isolation and characterization of bioactive compounds from plants' extracts. *Afr. J. Tradit. Complement. Altern. Med*, **8**: 1–10.

206. Thirumurugan, D., Cholarajan, A., Raja, S. S., and Vijayakumar, R. (2019). An introductory chapter: Secondary metabolites. Available online: https://www.intechopen.com/books/secondary-metabolites-sources-andapplications/an-introductory-chapter-secondary-metabolites.

207. Budisa, N. and Schulze-Makuch, D. (2014). Supercritical carbon dioxide and its potential as a life-sustaining solvent in a planetary environment communication. *Life*, **4**, 331–340.

208. Beckman, E. J. (2004). Supercritical and near critical CO_2 in green chemical synthesis and processing. *J. Supercritical Fluids*, **28**(2–3): 121–191.

209. Abbas, K. A., Mohamed, A., Abdulamir, A. S., and Abas, H. A. A. (2008). Review on supercritical fluid extraction as new analytical method. *Am. J. Biochem. Biotechnol.*, **4**: 345–353.

210. Herrero, M., Mendiola, J. A., Cifuentes, A., and Ibáñez, E. (2010). Supercritical fluid extraction: Recent advances and applications. *J. Chromatogr. A*, **1217**: 2495–2511.

211. Khaw, K. Y., Parat, O. M., Shaw, N. P., and Falconer, R. J. (2017). Solvent supercritical fluid technologies to extract bioactive compounds from natural sources: A review. *Molecules*, **22**: 1186.

212. Di Giacomo, G. (1989). Solubility of essential oils component in compressed SC-CO_2. *Fluid Phase Equil.*, **52**: 405–411.

213. Sodeifiana, G., Sajadiana, A. S., and Honarvarc, B. (2017). Mathematical modelling for extraction of oil from *Dracocephalum kotschyi* seeds in supercritical carbon dioxide. *Nat. Prod. Res.*, **32**: 795–803.

214. Baldino, L. and Reverchon, E. (2018). Challenges in the production of pharmaceutical and food related compounds by SC-CO_2 processing of vegetable matter. *J. Supercrit. Fluids*, **134**: 269–273.

215. Herrero, M., Cifuentes, A., and Ibañez, E. (2006). Sub- and supercritical fluid extraction of functional ingredients from different natural sources: Plants, food-by-products, algae and microalgae: A review. *Food Chem.*, **98**: 136–148.

216. Wang, L. and Weller, L. C. (2006). Recent advances in extraction of nutraceuticals from plants. *Trends Food Sci. Technol.*, **17**: 300–312.

217. Sato, T., Ikeya, Y., Adachi, S., Yagasak, K., Nihei, K., and Itoh, N. (2019). Extraction of strawberry leaves with supercritical carbon dioxide and entrainers: Antioxidant capacity, total phenolic content, and inhibitory effect on uric acid production of the extract. *Food Bioprod. Process*, **117**: 160–169.

218. Uquiche, E., Campos, C., and Marillán, C. (2019). Assessment of the bioactive capacity of extracts from *Leptocarpha rivularis* stalks using ethanol-modified supercritical CO_2. *J. Supercrit. Fluids*, **147**: 1–8.

219. Taylor, L. T. (1996). *Supercritical Fluid Extraction. Techniques in Analytical Chemistry*, pp 36–41, John Wiley & Sons, NY.

220. Klein, E. J., Náthia-Neve,s, G., Vardanega, R., Meireles, M. A. A., Antônio da Silva, E., and Vieira, M. G. A. (2019). Supercritical CO_2 extraction of α-/β-amyrin from uvaia (*Eugenia pyriformis* Cambess): Effects of pressure and co-solvent addition. *J. Supercrit. Fluids*, **153**: 1–6.

221. Wu, J., Liu, Z., Wang, B., and Pan, J. (2004). Measurements of critical parameters and saturation densities of dimethyl ether. *J. Chem. Eng. Data*, **49**(3): 704–708.

222. Taylor, L. T. (1996). *Supercritical Fluid Extraction: Techniques in Analytical Chemistry*, 1st edition, pp. 22, Wiley Inter Science.

223. Yousefi, M., Rahimi-Nasrabadi, M., Pourmortazavi, S. M., Wysokowski, M., Jesionowski, T., Ehrlich, H., and Mirsadeghi, S. (2019). Supercritical fluid extraction of essential oils. *Trends Anal. Chem.*, **118**: 183–193.

224. Herrero, M., Sánchez-Camargo, A. P., Cifuentes, A., and Ibáñez, E. (2015). Plants, seaweeds, microalgae, and food by-products as natural sources of functional ingredients obtained using pressurized liquid extraction and supercritical fluid extraction. *Trends Anal. Chem.*, **71**: 26–38.

225. King, J. W. and Friedrich, J. P. (1990). Quantitative correlations between solute molecular structure and solubility in supercritical fluids. *J. Chromatogr.*, **517**: 449–458.

226. Kanda, H., Fukuta, Y., Wahyudiono, and Goto, M. (2021). Enhancement of lipid extraction from soya bean by addition of dimethyl ether as entrainer into supercritical carbon dioxide. *Foods*, **10**: 1223. https://doi.org/10.3390/foods10061223.

227. Kanda, H., Hoshino, R., Murakami, K., Wahyudiono, Zheng, Q., and Goto M. (2020). Lipid extraction from microalgae covered with biomineralized cell walls using liquefied dimethyl ether. *Fuel*, **262**: 116590.

228. Girotra, P., Singh, S. K., and Nagpal, K. (2013). Supercritical fluid technology: A promising approach in pharmaceutical research. *Pharm. Dev. Technol.*, **18**: 22–38.

229. Martín, L., Marqués, J., González-Coloma, A., Mainar, A., Palavra, A., and Urieta, J. (2012). Supercritical methodologies applied to the production of biopesticides: A review. *Phytochem. Rev.*, **11**: 413–431.

230. Mattea, F., Martín, Á., and Cocero, M. J. (2009). Carotenoid processing with supercritical fluids. *J. Food Eng.*, **93**: 255–265.

231. Abimbola, T., Christodoulatous, C., and Lawal, A. (2021). Performance and optimization studies of oil extraction from *Nannochloropsis* spp. and *Scenedesmus obliquus*. *J. Cleaner Production*, **311**: 127295.

232. Herrero, M., Cifuentes, A., and Ibañez, E. (2006). Sub- and supercritical fluid extraction of functional ingredients from different natural sources: Plants, food-by-products, algae and microalgae: A review. *Food Chem.*, **98**: 136–148.

233. Belwal, T., Dhyani, P., Bhatt, I. D., Rawal, R. S., and Pande, V. (2016). Optimization extraction conditions for improving phenolic content and antioxidant activity in berberis asiatica fruits using response surface methodology (RSM). *Food Chem.*, **207**: 115–124.

234. Gañán, N. A., Dambolena, J. S., Martini, R. E., and Bottini, S. B. (2015). Supercritical carbon dioxide fractionation of peppermint oil with low menthol content: Experimental study and simulation analysis for the recovery of piperitenone. *J. Supercrit. Fluids.*, **98**: 1–11.

235. Yu, I. L., Yu, Z. R., Koo, M., and Wang, B. J. (2016). A continuous fractionation of ginsenosides and polysaccharides from *Panax ginseng* using supercritical carbon dioxide technology. *J. Food Process. Preserv.* **40**: 743–748.

236. Yoon, S. W., Pyo, Y. G., Lee, J., Lee, J. S., Kim, B. H., and Kim, I. H. (2014). Concentrations of tocols and gamma-oryzanol compounds in rice bran oil obtained by fractional extraction with supercritical carbon dioxide. *J. Oleo Sci.*, **63**, 47–53.

237. Duba, K. S. and Fiori, L. (2015). Supercritical CO_2 extraction of grape seed oil: Effect of process parameters on the extraction kinetics. *J. Supercrit. Fluids.*, **98**, 33–43.

238. Chavez-Gil, T. (2016). US-Patent #9,259,666-B1. Compact Extractor/ Separator Apparatus for Solid/Liquid/Gel Sample.

239. Chavez-Gil, T., Sobhi, H. F., and Pacheco-Laracuente, A. (2018). Improvements on semi critical assisted-solvent extraction of biocrude oil from *Spirogyra* sp. macro algae. *Am. J. Biochem. Biotechnol.*, **14**(2): 124–136. doi: 10.3844/ajbbsp.2018.124.136.

240. Chavez-Gil, T. (2019). A highly thermochemical assisted-solvent method for extraction of net proteins, enzymes, vitamins, antioxidants: Insights into step-by-step to fractionation to continuous extraction. *Proceedings 2019 TechConnet International Conference and Exposition.* Boston, MA, June 16–19.

241. Atkins, P. and De Paula, J. (2006). Molecular spectroscopy 3: Magnetic resonance, in *Physical Chemistry*, 8th edition, pp. 514, Freeman, Oxford University Press, UK.

242. NRC (National Research Council). Prudent Practices in the Laboratory. Handling and Management of Chemical Hazards. National Academy Press: Washington, DC, 2011.

243. Chavez-Gil, T. (2017). International Summit Fall 2017. Semi-critical high-performance extraction of polycyclic aromatic hydrocarbons: Improvements in decontamination of water, soil, and fuels raw material. *Energy Water Food Nexus*. Florida A&M University. Orlando, FL. Oct 18, **2017**.

244. Chavez-Gil, T. (2017). High yield/quality of net proteins, lipids, and antioxidants extracted through fractionation/one step chemical

method. *254th ACS-National Meeting & Exposition,* Aug 20–24, 2017. Washington, D.C, Agriculture & Food Division.

245. Maubois, J. L. (1984). Separation, extraction and fractionation of milk protein components. *Dairy Sci. Technol.*, **64** (645–646): 485–495.

246. Heck, J. M. L., Olieman, C., Schennink, A., van Valenberg, H. J. F., Visker, M. H. P. W., Meuldijk, R. C. R., and van Hooijdonk, A. C. M. (2008). Estimation of variation in concentration, phosphorylation and genetic polymorphism of milk proteins using capillary zone electrophoresis. *Intl. Dairy J.*, **18**: 548.

247. Greenfield, N. J. (2006). Using circular dichroism spectra to estimate protein secondary structure. *Nat. Protoc.*, **1**(6): 2876–2890. doi: 10.1038/nprot.2006.202.

248. Caroli, A. M., Savino, S., Bulgari, O., and Monti, E. (2016). Detecting casein variation in bovine milk. *Molecules,* **21**: 141. doi: 10.3390/molecules21020141.

249. Zhang, Y., Zhou, W., Yan, J-Q., Liu, M., Zhou, Y., Shen, X., Ma, Y-l., Feng, X-s., Yang, J., and Li, G-h. (2018). A review of the extraction and determination methods of thirteen essential vitamins to the human body: An update from 2010. *Molecules,* **23**: 1484. doi: 10.3390/molecules23061484.

250. Rowell, S., Oral, E., and Muratoglu, O. K. (2011). Annual Meeting of the Orthopaedic Research Society (ORS). Poster No. 1186.

251. Lee, J. Y., Yoo, C., Jun, S. Y., Ahn, C. Y., and Oh, H. M. (2010). Comparison of several methods for effective lipid extraction from microalgae. *Biores. Technol.*, **101**: 75–77. doi: 10.1016/j.biortech.2009.03.058.

252. Brennan, L. and Owende, P. (2010). Biofuels from microalgae: A review of technologies for production, processing and extractions of biofuels and co-products. *Energy Rev.*, **14**: 557–577. doi: 10.1016/j.rser.2009.10.009.

253. Elliott, D. C., Hart, T. R., Schmidt, A. J., Neuenschwander, G. G., and Rotness, L. J. (2013). Process development for hydrothermal liquefaction of algae feedstocks in a continuous-flow reactor. *Algal Res.*, **2**: 445–454. doi: 10.1016/j.algal.2013.08.005.

254. Ghasemi, N. F., Thomas-Hall, S. R., Durairatnam, R., Pratt, S., and Schenk, P. M. (2014). Comparative effects of biomass pre-treatments for direct and indirect transesterification to enhance microalgal lipid recovery. *Front Energy Res.*, **2**: 1–10. doi: 10.3389/fenrg.2014.00057.

255. Metherel, A. H., Taha, A. Y., Izadi, H., and Stark, K. D. (2009). The application of ultrasound energy to increase lipid extraction

throughput of solid matrix samples (flaxseed). *Prostaglandins Leukot Essent Fatty Acids*, **81**: 417–423. doi: 10.1016/j.plefa.2009.07.003.

256. Šoštarič, M., Klinar, D., Bricelj, M., Golob, J., and Berovič, M. (2012). Growth, lipid extraction and thermal degradation of the microalga Chlorella vulgaris. *New Biotechnol.*, **29**: 325–331. doi: 10.1016/j.nbt.2011.12.002.

257. Sahena, F., Zaidul, I. S. M., Jinap, S., and Karim, A. A. (2009). Application of supercritical CO_2 in lipid extraction: A review. *J. Food Eng.*, **95**: 240–253. doi: 10.1016/j.jfoodeng.2009.06.026.

258. Soh, L. and Zimmerman, L. (2011). Biodiesel production: The potential of algal lipids extracted with supercritical carbon dioxide. *Green Chem.*, **13**: 1422–1429. doi: 10.1039/c1gc15068e.

259. Huang, H., Yuan, X., Zeng, G., Wang, J., and Li, H. (2011). Thermochemical liquefaction characteristics of micro algae in sub- and supercritical ethanol. *Fuel Process. Technol.*, **92**: 147–153. doi: 10.1016/j.fuproc.2010.09.018.

260. Schafer, K. (1998). Accelerated solvent extraction of lipids for determining the fatty acid composition of biological material. *Anal. Chim. Acta*, **358**: 69–77. doi: 10.1016/S0003-2670 (97) 00587-4.

261. Long, R. D. E. and Abdelkader, E. (2011). Mixed-polarity azeotropic solvents for efficient extraction of lipids from nannochloropsis microalgae. *Am. J. Biochem. Biotech.*, **7**: 70–73. doi: 10.3844/ajbbsp.2011.70.73.

262. Kumar, R. R., Rao, P. H., and Arumugam, M. (2015). Lipid extraction methods from microalgae: A comprehensive review. *Frontiers Energy Res.*, **2**: 1–9.

263. Chavez-Gil, T., Sobhi, H. F., and Pacheco-Laracuente, A. (2018). Improvements on semi-critical assisted-solvent extraction of biocrude oil from *Spirogyra* sp. macro algae. *Am. J. Biochem. Biotech.*, **14**(2): 124–136. doi: 10.3844/ajbbsp.2018.124.136.

264. Moussa, T. A. A., Mohamed, M. S., and Samak, N. (2014). Production and characterization of di-rhamnolipid produced by pseudomonas aeruginosa TMN. *Brazilian J. Chem. Eng.*, **31**: 867–880. doi: 10.1590/0104-6632.20140314s00002473.

265. Yoon, M., Choi, J., Kim, G. H., Kim, D. H., and Park, D. H. (2013). Proteomic analysis of spirogyra variants mutant with high starch content and growth rate induced by gamma irradiation. *Bioprocess Biosyst. Eng.*, doi: 10.1007/s00449-013-0902-x.

266. Govindarajan, L., Senthilkumar, R., Raut, N., Bakheet, H., and Bakheet, F. J. (2010). Kinetic study of algae biomass grown in natural medium using spectroscopic analysis. *Algal Biomass Utilizat.*, **2**: 1–11.

267. Pacheco-Laracuente, A., Villarin-Ayala, S., and Chavez-Gil, T. (2014). High performance of algae-oil extraction: The effect of azeotropic mixtures on cell wall penetration. *Proceedings of the 247th Energy and Fuels ACS Spring Meeting* (ASM' 14), Dallas, TX, pp. 594–594.

268. Kongkiattikajorn, J. and Pongdam, S. (2006). Vanadium haloperoxidase from the red alga gracilaria fisheri. *ScienceAsia*, **32**: 25–30. doi: 10.2306/scienceasia1513-1874.2006.32(s1). 025.

269. Shimamoto, G. G., Favaro, M. M. A., and Tubino, M. (2015). Simple methods via mid-IR or 1H NMR spectroscopy for the determination of the iodine value of vegetable oils. *J. Braz. Chem. Society*, **26**: 1431–1437.

270. Hema, R., Kumaravel, S., Gomathi, S., and Sivasubramaniam, C. (2010). Gas chromatography-mass spectroscopic analysis of lawsonia inermis leaves. *New York Sci. J.*, **3**: 141–143.

Index